PRENTICE-HALL

Foundations of Cultural Geography Series

PHILIP L. WAGNER, *Editor*

RICHARD E. DAHLBERG, *Series Cartographer*

* *In Prentice-Hall's Foundations of Economic Geography Series, also.*

Foundations of Cultural Geography Series

RICHARD E. DAHLBERG, *Series Cartographer*

Geography

of

Domestication

ERICH ISAAC

City College of New York

PRENTICE-HALL, INC., Englewood Cliffs, N.J.

Maps produced by the author.

PRENTICE-HALL INTERNATIONAL, INC., London
PRENTICE-HALL OF AUSTRALIA, PTY. LTD., Sydney
PRENTICE-HALL OF CANADA, LTD., Toronto
PRENTICE-HALL OF INDIA PRIVATE LTD., New Delhi
PRENTICE-HALL OF JAPAN, INC., Tokyo

Foundations of Cultural Geography Series

The title of this series, Foundations of Cultural Geography, represents its purpose well. Our huge and highly variegated store of knowledge about the ways that humans occupy and use their world becomes most meaningful when studied in the light of certain basic questions. Original studies of such basic questions make up this series of books by leading scholars in the field.

The authors of the series report and evaluate current thought centered on the questions: How do widely different systems of ideas and practice influence what people do to recreate and utilize their habitats? How do such systems of thought and habitat spread and evolve? How do human efforts actually change environments, and with what effects?

These questions are approached comparatively, respecting the great range of choice and experience available to mankind. They are treated historically as well, to trace and interpret and assess what man has done at various times and places. They are studied functionally, too, and whatever controlling processes and relationships they may reveal are sought.

Diverse tastes and talents govern the authors' attack on these problems. One deals with religion as a system of ideas both influencing and reflecting environmental conditions. Another evaluates the role of belief and custom in reshaping plant and animal species to human purposes. Some consider the use and meaning of human creations, like houses or cities, in geographic context; others treat the subtle and complex relationships with nature found in agricultural systems of many sorts. One author looks at an entire country as a culturally-shaped environment; another analyzes the mechanics of the spread of customs and beliefs in space. All work toward an understanding of the same key problems. We invite the reader to participate actively in the critical rethinking by which scholarship moves forward.

PHILIP L. WAGNER

Contents

to the memory of my son
Noah
11·2·62 — 5·7·68

The forms and distributions of living organisms that man has shaped out of nature to his own use record an otherwise forgotten history. More crucial than ever to human livelihood, domestic plants and animals arrest geography's interest not only because of their obvious utility, but probably even more as indicators both of past migrations and connections, and of modes of the relationship between environments and human groups. Indeed, as Professor Isaac amply documents, motives other than utility inspired domestication. The geography of domestic plants and animals, elucidated by the bits of clues contributed by many disciplines, gives testimony still to sentiments and ideals of earlier mankind.

PHILIP L. WAGNER

introduction

Where and how plants and animals were first domesticated is a question that has consistently proved fascinating to students of many disciplines. Since the events being studied lie in the mists of prehistory, it is not surprising that the question has never been answered to everyone's satisfaction. While archaeological materials bearing on the question each year grow more numerous, the problem of their interpretation continues to be perplexing. There may well never be definitive answers to the crucial questions: Where did domestication first arise? Did it spread from a single source or was it discovered independently in many places? What plants and animals were first domesticated? What people or peoples first domesticated?

These questions remain so intriguing to scholars because of the crucial importance domestication has had for human history. Domestication was the invention that made possible the development of pastoral and agricultural economies and therefore made populous and complex human societies viable. It has thus proved to be the single most important intervention man has ever made in his environment. As philosopher Eric Hoffer has noted, "Technologically, the Neolithic Age lasted even in Western Europe down to the end of the eighteenth century." [1] Thus the drama of the achievement is heightened because man's inventiveness in this area was largely exhausted with the earliest domestication. Only in our own time have alternative sources for some of the needs served until now by domestic plants and animals been developed. The most striking achievements thus far are in synthetic fibres, dyes and drugs, and there is the potential, of course, for revolutionary new sources of food.

In approaching the problem of the origin of domestication, no student

[1] Eric Hoffer, *The Ordeal of Change* (New York: Perennial Library, Harper & Row, Publishers, 1967), p. 48.

1

of the subject can avoid working within the framework of the known geographic relationships of food producing economies (i.e., economies based on the systematic use of plant and animal domesticates) in the Old World. New World societies may have domesticated plants independently or under the stimulus of the Old World—scholarly opinion diverges on this point. Some plants in ancient times may have been introduced from the New World to the Old. Until more is known about prehistoric cultural movements between the two no more than speculation is possible. In any event, domesticating societies in the Americas are apparently younger than the oldest known in the Old World.

As a result of the age of discoveries, plant and animal introductions after 1500 A.D. made a radical impact; the starting point therefore becomes the Old World distribution of economies prior to 1500 A.D.[2] That distribution in generalized outline involved a zonal pattern starting in the north with circumpolar hunters. Pastoral nomads bordered them in the south, and they in turn gave way to a zone of peasant economies. South of the plow using peasantries was another pastoral nomadic zone with enclaves of primitive food collectors and hunters. Finally came (and indeed comes) the zone of tropical planters, again interspersed with food collecting and hunting groups. Of the economies found within these zones, those of pastoral nomads, peasants, and planters was based on domestic plants and animals.

The questions which immediately occur are:

1. What is the relationship of the entire complex of food producing cultures to the various hunting and gathering societies?
2. In which of these generalized zones did the domestication of plants or animals originate, if indeed domestication was a single invention?
3. To what extent are the economies of the different zones related?

[2] Of course there were many plant and animal transfers prior to 1500 A.D., but the scope of introduction vastly accelerated after the age of discoveries.

CHAPTER 1 *major theories concerning the origins of domestication*

The traditional view of the origin of agriculture was an uneasy mixture of cyclical–developmental doctrine and environmental determinism. The Greeks combined both views. On the one hand they believed that human economy had developed through three stages: first came a hunting and gathering stage; this slowly led to the domestication of animals and a pastoral nomadic stage; finally came the invention of agriculture. This three stage theory has been attributed by various writers of antiquity to Dicaearchus (ca. 320 B.C.), but it had probably been formulated considerably earlier. On the other hand, the Greeks noted that the zonation of economies conforms in its broad outline with the zonation of climates. The fact of this correspondence led to the assumption that physical conditions adequately explained the variety of food producing societies, e.g., where there is steppeland, it was presumed that pastoral nomadism inevitably developed. The two views were sometimes combined: economies passed through stages, but physical conditions determined which stage would persist in what area. Analogous conceptions are found in Indian and other Asian literatures. Even in the biblical narrative, there is a notion of a sequence of stages in the development of civilization, although according to Genesis IV, nomadic pastoralism appears later than cultivation.

These ancient views persisted in Europe throughout the Middle Ages and in different, this time evolutionary, form into modern times. However the modern version postulated a unilinear sequence from "lower" to "higher" stages, whereas the Greeks had combined a stage theory with a general cyclical view of history. Originally there had been a Golden Age when an exclusively plant eating man lived in harmony with himself and animals (this conception is of course characteristic of many cultures). From this stage there could only be devolution: there was certainly no notion of a steady progress towards anything. Eventually man would return to the beginning to start all over again.[1] From this point of view there is not a direct line from the Greeks to the modern evolutionary approaches of geography or ethnology. The Greek classification carried over into modern times and the categories may thus sound similar, but the underlying conception of the processes and direction of history are different. In Europe in the eighteenth century the evolutionary theory was dominant, the historical school trying to account for the differences between peoples in different areas of the world by assigning them to different stages on parallel, independent, but essentially identical ladders of evolution. In a recent study Fritz L. Kramer shows how little the views of nineteenth century philologists, economists, and anthropologists had changed, even though some of them depart on details of the sequences and number of stages.[2]

The stage theory in its evolutionary form was first challenged toward the end of the eighteenth century, when it suffered through a process of attrition. The complete sequence, it was noted, held only for the temperate zone, and hunting and herding were eliminated as stages in the torrid zone. Environmental determinism, as with the Greeks, became combined with the basic theory, and it was assumed that different economies evolve through stages appropriate to the climates in which they are found. Perhaps the rationalistic conception of man was influential in the acceptance of the notion of environmental determinism, for all men were regarded as fundamentally the same, and the differences between their societies could be explained by the differences in the environments in which they found themselves.[3]

From Lord Kameshto Alexander von Humboldt to Friedrich Ratzel to Eduard Hahn, there is an increasingly sophisticated awareness of the failure of the notion of a universal sequence to explain the geographic facts of domestication. They noted that nomadic herding is not a necessary intermediate step between hunting and farming, for nowhere in the Western hemisphere did herding people emerge from hunters in spite of an abundance of domesticable herd animals. They saw that the only herding complexes in the Old World were those contiguous to settled peoples, and where there were hunters without farmers, as in Australia, no herding complexes developed. Conversely, they noted, there were people who derived part of their sustenance from collecting and hunting and part from cultivating the soil, without having domesticated herd animals.

What led to these observations was the attempt to map human econo-

---

[1] On cyclical history see Joseph Maier, "Cyclical Theories," in *Sociology and History,* eds. Werner J. Cahnman and Alvin Boskoff (New York: The Free Press, 1964), 41–62.
[2] "Eduard Hahn and the End of the 'Three Stages of Man,'" *The Geographical Review,* LVII, No. 1 (1967), 73–89.
[3] Concerning the influence of the natural sciences and Greek "evolutionism" on formerly prevailing doctrines of multilinear and parallel evolution of cultures see Kenneth E. Bock, *The Acceptance of Histories: Toward a Perspective for Social Science,* University of California Publications in Sociology and Social Institutions, 3, No. 1 (Berkeley and Los Angeles: University of California Press, 1956).

mies and the effort to produce a general human and cultural geography. One result of this effort was that the traditional and vague categories such as agriculture, hunting and herding broke down. In itself this increasing differentiation did not at first threaten the evolutionary view. The sequences were simply modified and adjusted to various habitats. For example, even when "agriculture" was refined as a category, the different complexes of food production found in different parts of the world were treated as parts standing for the whole. Richard Braungart in 1881 distinguished between plow agriculture and agriculture without plow, but then propounded the conventional three stages.[4] Agriculture might mean one thing in one place and something else in another, but it still followed upon nomadic pastoralism which in turn derived from hunting and collecting. Nonetheless all attempts at elaboration suffered from the fact that similar habitats had clearly failed to produce similar adaptations, and quite different environments sometimes permitted quite similar cultures to thrive. Furthermore, prehistory and ethnography, which developed rapidly in the nineteenth century, supported the idea that the exotic primitives of the period were similar in their technology to the populations of Europe in prehistoric times. An alternative explanation for the distribution of world economies gained fashion. Perhaps not parallel development, but diffusion was the key.[5]

It is difficult to understand why the notion of diffusion should have taken hold so strongly, for the attraction of the evolutionary idea is enormous. In addition, on the basis of the very widespread conviction of the psychic unity of mankind, it could equally well have been argued that the sequences were there, but were different from those originally postulated. The notion of diffusion, of course, is not a new one. In occidental thought the conception of a human cradle from which mankind spread under the guidance of providence was old and attempts at writing universal history in this vein, with or without providence, were made. Thus Johann Gottfried Herder and others looked at the Near East as the cradle of civilization, and therefore of agriculture. Prehistory and new linguistic methods seemed to support diffusionist interpretations, although often with a biblical bias. The development of the European colonial empires and the attendant migrations and introduction of European plants, animals, and agricultural systems may also have contributed. The world was rapidly changed through transmission not only between metropolitan countries and their colonies but also between the colonies themselves. It seemed only reasonable to suppose that migration, dispersal of techniques, and intellectual exchange in the past had played a similar if hitherto neglected role. Moreover, for all the technological revolutions of nineteenth century Europe, there was a sudden suspicion of the limita-

[4] Die Ackerbaugeräthe in ihrer praktischen Beziehungen wie nach ihrer urgeschichtlichen und ethnographischen Bedeutung (Heidelberg: C. Winter, 1881), p. 4
[5] In addition the geographers laid the foundations of central European scientific ethnology. See Robert Heine-Geldern, "One Hundred Years of Ethnological Theory in the German-Speaking Countries: Some Milestones," *Current Anthropology*, 5, No. 5 (1964), 407–16, and the "Comment" by Paul Leser, *ibid.*, 416–17.

tions of man's innate inventiveness, a suspicion that seems to have struck geographers early and with particular force. Modern social–evolutionary theory postulated a uniform potential for human inventiveness called forth by needs and realized in technological achievements that build upon one another. Yet such seemingly simple inventions as the wheel and the plow, the geographers noted, were not invented everywhere they might have been useful. Perhaps, they concluded, there have been centers of invention from which the inventions themselves, with their tremendous power for transforming human society, were distributed. The study of domestication and of the origins of agricultural systems led geographers of many specialties to the nearly century old preoccupation with dispersals and diffusion, a concern that has made them an irritant, although a valuable one, to the anthropological, sociological, and economic schools that have emphasized the function of cultural elements in explaining the presence of domestic animals and plants within a culture, as indeed the presence of any cultural elements.

Still another shift in perspective had important consequences for the views held concerning plant and animal domestication. German geographers in particular had characterized agricultural complexes by their main implements. Tools were ranked on an evolutionary scale and associated farming systems dated in accordance with the tools they had used. This preoccupation with implements derived from, and at the same time reinforced the conviction that agricultural inventions were rational responses to practical needs. But a profound change in this way of thinking came when the focus was placed on the plant and animal domesticates that were characteristic of these complexes. Suddenly it appeared that utilitarian motivation could not account for important domestications. How would it have been possible, for example, to foresee in advance of its domestication that the urus, a fierce and dangerous animal, would become the tame meat providing, milk producing power source that domestic cattle became? Once the focus was on the domesticates themselves innumerable such problems appeared. Nonutilitarian motives for domestication had to be explored, and this in turn had an influence upon views of the nature of man in an era dominated by rationalistic conceptions, whether in philosophy or the natural and social sciences.

All the new ways of thinking about domestication were best crystallized in the writing of Eduard Hahn, whose work was decisive in demolishing the stage-theory which had borne up for millennia.[6] Friedrich Ratzel before him had suggested that domestication was the result of a protective attitude toward potential domesticates, an attitude that can be fully developed only among sedentary peoples.[7] It followed that herd animal and plant domestication were both achievements of sedentary farmers and that pastoral nomadism was a later development. Hahn

[6] Eduard Hahn's most important works are: *Demeter und Baubo; Versuch einer Theorie der Entstehung unseres Ackerbaus* (Lübeck: in commission bei Mat Schmidt, 1897); *Die Haustiere und ihre Beziehungen zur Wirtschaft des Menschen* (Leipzig: Duncker & Humblot, 1896); *Die Entstehung der Pflugkultur* (Heidelberg: C. Winter, 1909); *Von der Hacke zum Pflug* (Leipzig: Quelle & Meyer, 1914).
[7] *Völkerkunde* (Leipzig: Bibliographisches Institut, 1886), I, 57.

believed that the motive for the domestication of herd animals and especially cattle was religious. He pointed out that the usefulness of wild cattle for labor or milk could not have been foreseen until the animals were in fact domesticated.

Hahn further showed that the sequence hunting–herding–agriculture cannot be maintained.[8] Wild rice collecting and hunting in the Great Lakes area of North America went hand in hand with plant cultivation, while the herding of domesticated animals was unknown. Hunting and collecting people thus seemed to be able to pass directly to hoe cultivation without passing through an animal domesticating stage. Whether hunters became hoe cultivators in one single hearth from which this form of cultivation then spread, or independently in a number of locations is left an open question. For grain domestication, the question remained whether cereal growing, which is so closely associated with plow agriculture, is of independent origin or derives from hoe cultivation, which seems closely associated with subtropical and tropical root crops. Hahn believed that the hoe was used before the plow in cereal growing, and that in Babylonia, fields in which barley and wheat were domesticated were cultivated with hoes. In addition, he felt that the survival of the hoe within the plowing realm showed that it continued to fulfill functions from which plows were unable to displace it.

In short, Hahn believed that hoe cultivation underlies all food producing complexes, and that the plow evolved from the hoe. Hahn did not take a stand on where plant domestication originated. Hoe cultivation is characteristic mainly of tropical areas, but since in his view the hoe preceded the plow (and, in fact, underlies all agriculture) plant domestication may have started in the classical areas of plow civilization. The hoe's first object might have been wild grasses rather than the tuberous or root crops that are often considered characteristic of hoe agriculture when the term is used today. The realm of plow farming appeared as a localized development in the center of the broader hoe world. Pastoral nomadic realms appeared as secondary offshoots of grain growing and herd animal domesticating plow-peasants.

Hahn did more than demolish the stage theories and provide a reconstruction of agricultural history. He examined the historical derivation of agricultural implements as a guide to the history of domestication and tried to show that a new world picture might serve as the basis for economic innovations. Beyond this, his sudden examination of previously unexamined assumptions was to open the way to new studies. But while Hahn opened the way to new examinations of agricultural origins, his specific theses and conclusions were pushed aside by new ethnographical data and new ethnological and sociological interpretations.

As early as 1861, Johann Jakob Bachofen, a Swiss jurist, exerted a powerful influence on ethnology by his thesis of a matriarchal stage of mankind which preceded the patriarchal societies of antiquity. In his

---

[8] On the evolution of Hahn's thought concerning these "economic forms" and their subsequent modification by others, see Kramer, *op. cit.*, pp. 82–89.

great work *Das Mutterrecht* he stressed that originally women tilled and harvested the land.[9] Culture–historical ethnologists who followed him theorized that the hypothetical female domesticator was a food gatherer who took the first steps toward plant domestication when she protected certain plants in her range. In 1909 the ethnologist Fritz Graebner argued that while the systematic care of plants probably originated in a planting complex dominated by women, herd animal domestication was an achievement of male hunters. The association of animal and plow came about when these different cultures touched and mingled.[10]

Later ethnologists elaborated the hypothetical transition by which women changed from food foragers to planters. Kaj Birket-Smith, a Danish ethnologist, illustrated the process by pointing to western Australian female yam collectors who bury wild yams after they have dug up the edible tubers.[11] Others suggested such assumed stimuli as settlement-associated refuse heap fauna that female food gathers who revisited former settlement sites presumably observed to be particularly luxuriant. Of course, what was most stressed was that men have generally remained hunters in such food collecting groups.[12]

Although in principle the transition from collecting to planting could have occurred anywhere in the Upper Paleolithic period, scholars more and more tended to assume that it actually occurred in the tropics or specifically in the wet-and-dry tropics, for no apparent reason except that the pattern had been reconstructed on the basis of tropical evidence. Thus southeast Asia, including the archipelagoes, was ever more firmly fixed as the source region for the initial domestication of plants. Tuberous plants and clones (e.g., yams, taro) and trees on which particular care is lavished and which are often propagated by planting (e.g., sago-palm, banana) in this area were considered to be the first domesticates. These cultures were also held to have domesticated the smaller domestic animals, mainly pigs and fowl. This thesis was developed most consistently by Father Wilhelm Schmidt who advanced the theory that a nomadic–patriarchal culture in the steppes of inner Asia developed from a hunting culture which had once stretched across the northern regions of Eurasia. He considered various reindeer herding peoples in north-central Asia, who often only herd semidomestic animals, as prime ethnological examples of the transition from the hunting of herd animals to their domestication. In Schmidt's view, the successful domestication of reindeer led some groups to apply similar or derived techniques to domesticate other herd animals, whose range was farther south. It was the meeting of the

[9] J. J. Bachofen, *Das Mutterrecht: Eine Untersuchung über die Gynaikratie der alten Welt nach ihrer religiösen und rechtlichen Natur* (Stuttgart: Verlag Krais und Hoffmann, 1861).

[10] F. Graebner, "Die melanesische Bogenkultur und ihre Verwandten," *Anthropos*, 4 (1909), 1030.

[11] Kaj Birket-Smith, *The Paths of Culture*, trans. Karin Fennow (Madison and Milwaukee, Wis.: The University of Wisconsin Press, 1965), p. 153.

[12] Richard Thurnwald, *Die menschliche Gesellschaft in ihren ethno-soziologischen Grundlagen* (Berlin & Leipzig: W. de Gruyter & Co., 1931), I, 94. Ferdinand Herrmann, "Die Entwicklung des Pflanzenanbaues als ethnologisches Problem," *Studium Generale*, 11, No. 6 (1958), 352–55.

patriarchal nomadic complex with matriarchal planting cultures which gave rise to a plow–cattle–cereal peasantry.

Schmidt elaborated further on the role of women in domestication. He believed that the historical transition from unspecialized collecting to female domesticating cultures was via an intermediate stage described as hunter–planter (Jäger–Pflanzer) culture. This culture differed from earlier food collecting cultures in the presence of clan-totemism and the more important role of women. With the increasing importance of plant food and incipient domestication, male hunting declined and became accessory to the main food producing functions of women. Men often specialized in fishing which did not interfere with the sedentary life that female planting demanded. Women used the grubbing stick, which with at most a few modifications, became the digging stick of many tropical cultivators.[13]

An adequate critique of the contribution of Schmidt and others who worked in this tradition cannot be given in the frame of this study. Suffice it to say that the relation of totemism to the world picture of primitive collectors is an uncertain one, and that "totemism" itself as a religious and ethnographic category has been under attack for many years. It should also be noted that the position of men in primitive tropical systems of agriculture and in recent or contemporary food collecting societies has been underestimated. In fact, Father Schmidt's reconstructions of cultural complexes, including the patriarchal and matriarchal primary domesticating cultures, have been largely abandoned by ethnology. Thus a major underpinning of Schmidt's revision of Hahn's thesis has been lost.

Nonetheless, important elements of Schmidt's culture–historical arguments have been retained by recent scholars of agricultural origins, perhaps the most influential of whom is the geographer Carl O. Sauer. Sauer placed new emphasis on the attempt to find the geographic hearth of domestication of both plants and animals. Although the possibility of recurrent or independent domestication of individual domesticates is not denied, there is a sharpened awareness of the tremendous accomplishment that domestication constitutes. The domestication of different plants and animals at different times and places then becomes incidental to the first great breakthrough involved when the possibility of domestication was grasped and implemented. In essence what the "geographic school" in this area has been trying to do is to specify the biological, social and environmental conditions in which the original transition from Upper Paleolithic specialized hunting and gathering to food production could have occurred, and then find the area where the specifications are best met. In many respects Schmidt and some other ethnologists fit into this group, but they do not focus as sharply on a geographic hearth and on the actual potential domesticates as do the geographers. A tremendous

---

[13] Wilhelm Schmidt and Wilhelm Koppers, *Völker und Kulturen; Gesellschaft und Wirtschaft der Völker* (Regensburg: Verlag Josef Habbel, 1924), pp. 194–224, 502–38; Wilhelm Schmidt, "Zu den Anfängen der Herdentierzucht," *Zeitschrift für Ethnologie,* 76 (1951), 1–41; 201–4; Fritz Bornemann, "P. W. Schmidts Studien über den Totemismus in Asien und Ozeanien," *Anthropos,* 51 (1956), 595–794.

impetus for this type of analysis came from the work of the great Russian plant geographer N. Vavilov who, on the basis of extensive field work, outlined centers of variability of plants and concluded that domestication must have occurred within these centers. This in turn raised the question of which plants were first domesticated in which center, and which center is the oldest.

Sauer provides a number of criteria which must be met by the area in which domestication first occurred and by the peoples who first domesticated.

1. Plant domestication could not occur in an area where people were in chronic need of food. People threatened by famine cannot afford the leisurely experimentation that would lead to more abundant food in the distant future.
2. The hearth must be in an area of great diversity of plants and animals, "where there was a large reservoir of genes to be sorted out and recombined." The implication is for a diversified terrain with a variety of climates.
3. Domestication could not first occur in large river valleys which require advanced works for water control.
4. Cultivation had to begin in wooded lands, for while primitive man "could readily open spaces for planting by deadening trees," he did not have the capacity to break the sod of grasslands.
5. The earliest farmers must have had skills that predisposed them to farming. Sauer thinks hunters could not have been these people, but "wood dwelling" ax users must have been the ancestors of domesticators.
6. Finally, and above all, the original domesticators had to be sedentary since "what is food for man is feast for beasts," and if men were not there permanently to watch over their growing crops, there would be no crops to harvest.[14]

On the basis of these criteria Sauer concludes that the hearth of plant domestication is Southeastern Asia. The domesticators were Mesolithic fishing folk living along fresh water, the location which supplied them with staple food and made sedentary life possible. They made use of multipurpose plants which provided starchy food, substances for toughening nets and lines, and drugs and fish poisons. And while the men fished and built boats, the women cultivated and cooked. For Sauer, the characteristic form of monsoon land agriculture which multiplies food by the use of clones is evidence to support his view of the precedence of this kind of agriculture. Individual plants are chosen and propagated by division, and this practice leads to the rise of an "extraordinary lot of forms that are completely dependent on man for their existence." Many have lost the ability to produce seeds: "break the continuity of this operation and the plant may be lost."[15] These plants provide mainly carbohydrates, but the required supplementary foods were available in fish and shellfish.

According to Sauer the art of plant domestication spread north in

[14] Carl O. Sauer, *Agricultural Origins and Dispersals* (New York: The American Geographical Society, 1952), 20–22.
[15] *Ibid.*, p. 25.

the Old World and perhaps from the Old World to the New, although, like Hahn, Sauer leaves open the possibility of independent centers of domestication in tropical areas of the New World. In the course of its northward expansion the art of cereal cultivation would have developed. Perhaps the practice of planting rice in seedbeds and then transplanting individual seedlings, although not essential to the successful cultivation of rice, signifies the persistence of planting habits appropriate to clones.

The Mesolithic fishing folk who domesticated plants, Sauer believes, domesticated dogs and pigs by making them pets and adopting them into the human family. Fowl, he thinks, were domesticated for ceremonial purposes.[16] The herd animals—cattle, sheep, and goats—were domesticated by cereal farmers in southwest Asia, themselves derived from or influenced by the earlier Old Planter complex of southeast Asia. Sauer's position is thus in the culture–historical tradition of Graebner and Schmidt and in accord with Hahn in asserting that some domestications were religious in origin.[17]

It took roughly a century, one that began with Bachofen and ended with the cultural geographers of the present day, to bury the classical stage theory and to put the evolutionary theory into eclipse if not oblivion. There is today no view of the origin and progress of plant and animal domestication that commands the universal assent once given to the theory of the three stages of man. The only real agreement that can be found today among geographers and culture–historical ethnologists is that the domestication of plants preceded that of herd animals, and that the original domestication of plants took place in a region (generally held to be Southeast Asia) different from the area where the big herd animals were first domesticated (the Near East or western Asia).

Indirectly the thesis of a tropical origin of plant domestication has led to a reformulation of the "Golden Age." This concept survived the end of the stage theory on the periphery of a romantically tinged scholarship. The first domestication of plants was held by many scholars to be in forested areas where a vast range of vegetarian food was available and in which man's closest relatives, the vegetarian anthropoid apes, are found. It was not difficult for vegetarians and tree enthusiasts to construct hypotheses of an original plant domesticating culture characterized by blissful innocence, knowing neither war nor meat. From cherishing the arboreal diet that nourished them, they took the first steps to domestication by propagating the trees that gave them food.[18]

---

[16] Sauer, *Agricultural Origins,* 86–90.

[17] Other geographers have substantially accepted Sauer's reconstruction. Cf. Hermann von Wissmann, "On the Role of Nature and Man in Changing the Face of the Dry Belt of Asia," in *Man's Role in Changing the Face of the Earth,* ed. William L. Thomas, Jr. (Chicago: The University of Chicago Press, 1956), pp. 283–87, and "Ursprungsherde und Ausbreitungswege von Pflanzen und Tierzucht und ihre Abhängigkeit von der Klimageschichte," *Erdkunde,* XI, No. 2 (May 1957), 81–94, and XI, No. 3 (August 1957), 175–93.

[18] On the history of vegetarian ideas see Arthur O. Lovejoy and George Boas, *Primitivism and Related Ideas in Antiquity* (Baltimore, Md.: Johns Hopkins University Press, 1935) I, and Russell Lord, *The Care of the Earth, A History of Husbandry* (New York: Mentor Book by The New American Library, 1963), pp. 58–70.

But as anthropological studies brought an ever widening knowledge of primitives and of prehistory, the notion of an innocent peaceful early society became difficult to sustain. The hypothetical age had to be pushed so far into the past that it became another Golden Age myth. Thus Henry Bailey Stevens pushes the origins of plant domestication almost as far back as the transition from the anthropoid state to humanity, and makes horticulturists the first domesticators. His fascinating book *The Recovery of Culture* is a combination of a vegetarian culture-history and a utopian recovery of man's earlier vegetarian values.[19]

Whatever the value of such theses in a broader sense may be, they do serve a useful purpose in pointing to the importance of tree culture, which is often neglected by students of domestication. In Southeast Asia and the Pacific today there are food gathering and hunting societies which depend to a significant extent on tree crops (e.g., sago-palms, *Sagus* spp.), protect the trees, and engage in a rudimentary type of propagation. Tree domestication can also be observed today among some New World farming societies. Carl L. Johannesen illuminates the ongoing process of tree domestication in his studies of the pejibaye palm, *Bactris gasipaes*, in Central America.[20] Of course none of the contemporary tree domesticators can be equated with the ancient domesticating cultures, for they have all been influenced by older farming centers.

Apart from archaeologists, the only major recent exponent of a theory that locates the oldest domestication outside the tropics is the prehistorian V. Gordon Childe, whose riverine–oasis hypothesis has received wide attention. The hypothesis states that postglacial desiccation forced man and animals into symbiotic associations in river valleys or oases of the Near East. From such close association of men and domesticable animals domestication inevitably proceeded.[21] But while there is evidence that during the last pluvial (which corresponds approximately to the Würm glaciation in Europe) there was more rainfall in North Africa than at present, the evidence from southwestern and western Asia suggests that late Pleistocene or post-Pleistocene climates were substantially modern in those areas. Charles A. Reed warns against the tendency of prehistorians "to enfold southwestern Asia into the conclusions reached

[19] Henry Bailey Stevens, *The Recovery of Culture* (New York: Harper & Brothers Publishers, 1949).

[20] Carl L. Johannesen, "Domestication Processes used by Subsistence and Small Scale Farmers" (Mimeo., Discussion Papers in Cultural Geography, 61st Annual Meeting of the Association of American Geographers, Columbus, Ohio, April 19, 1965), 38–47; also, "The Domestication Processes in Trees Reproduced by Seed: The Pejibaye Palm in Costa Rica," *The Geographical Review*, LVI, No. 3 (1966), 363–76.

[21] V. Gordon Childe, *The Most Ancient East* (London: Routledge & Kegan Paul, Ltd., 1929), pp. 42–46. Earlier versions of the same thesis are found in R. Pumpelly, "Explorations in Turkestan: Expedition of 1904: Prehistoric Civilizations of Anau" (Washington, D.C.: Publications of the Carnegie Institute, 1908), I, 65–66, and in H. J. Peake and H. Fleure, *Peasants and Potters* (Oxford: The University Press, 1927), III, 14. For a critique of such theses including Arnold J. Toynbee's version of the riverine-oasis hypothesis in his *A Study of History*, 2nd. ed. (London: Oxford University Press, 1935), I, 305, see Karl W. Butzer, "Environment and Human Ecology in Egypt During Predynastie and Early Dynastic Times," *Bulletin de la Société de Géographie d'Egypte*, XXXII (1959), 78–85.

on the basis of North African data. . . ."[22] Even in North Africa the process of desiccation was fluctuating, and the only period when desiccation had proceeded to the point where the close association described by Childe must have occurred was more than 2000 years before the earliest evidence of domesticated animals in the region. Evidence for a way of life that included domesticated animals and plants in North Africa does not in fact appear until the onset of the "Neolithic Wet Period." The earliest known site is in the Fayum ca. 4200 B.C., which suggests that the introduction of domesticates into Africa from the east was facilitated by increased humidity. Recent studies by Shalem and others show that climatic drying up affected the Near East much less than lower latitude deserts.[23] Given the type of relief found in the Levant and the Zagros-Taurus arc, desiccation would not have destroyed life zones, but merely pushed them farther up into hills and bordering ranges. Where elevations were not sufficiently high, the vegetation would have died out, and with that vegetation the animals dependent on it. "There would be no 'crowding-down' into river valleys and oases of animals which—contrarily—could only, however slightly in each generation, be living at even higher elevations." As Reed points out, even if some animals had made it to oasis or river bank environments, preagricultural man in a deteriorating environment would have killed anything he could. He "would be an exterminator, not a conservator via domestication."[24] Further, many detailed studies have shown that semiaridity in the Near East is a consequence rather than a precondition of domestication. Goats, in particular, destroy leaves, saplings, young shoots, and branches of trees and bushes and thus prevent the regeneration of forest. Finally, the usefulness of climatic change in explaining the origin of domestication has recently been challenged by C. Vita-Finzi, who argues that geomorphical conditions may be as relevant as climate, and sometimes more so.[25]

[22] Charles A. Reed, "A Review of the Archaeological Evidence on Animal Domestication in the Prehistoric Near East" in *Prehistoric Investigations in Iraqi Kurdistan,* by Robert J. Braidwood and Bruce Howe (Chicago: The Oriental Institute of the University of Chicago, University of Chicago Press, 1960), Studies in Ancient Oriental Civilization, No. 31, 122, henceforth cited as "Archaeological Evidence." For an excellent discussion on climatic conditions in the Mediterranean region and eastern Africa in the late Pleistocene and the bearing of climate on domestication see Karl W. Butzer, *Environment and Archaeology* (Chicago: Aldine Publishing Company, 1964), pp. 285–316, 425–26, and also Butzer's earlier "Quarternary Stratigraphy and Climate in the Near East," *Bonner Geographische Abhandlungen,* No. 24, Geographisches Institut der Universität Bonn (Bonn: Ferd. Dümmler Verlag, 1958), especially pp. 108–18.

[23] N. Shalem, "La Stabilité du climat en Palestine," Research Council of Israel, Special Publication II (1953), 153–75. Herbert E. Wright, Jr., "Climate and Prehistoric Man in the Eastern Mediterranean," in Braidwood and Howe, *Prehistoric Investigations,* pp. 71–97. Hans Bobek, "Klima und Landschaft Irans in vor- und frühgeschichtlicher Zeit," *Geographischer Jahresbericht aus Österreich,* XXV (1957), 1–42. Robert L. Raikes and Robert H. Dyson, Jr., "The Prehistoric Climate of Baluchistan and the Indus Valley," *American Anthropologist* (1961) 63, No. 2, Part I (1961), 265–81.

[24] Reed, "Archaeological Evidence," p. 123.

[25] C. Vita-Finzi, "Geological Opportunism." In Peter J. Ucko and G. W. Dimbleby, eds., *The Domestication and Exploitation of Plants and Animals* (Chicago: Aldine Publishing Company, 1969), pp. 31–34.

The inability to obtain clear proof for any theory of agricultural origins has led some surgical temperaments to cut through the knots of suppositions and hypotheses. Such temperaments are represented in all fields concerned with domestication. The "accident theory" is the oldest expression of this "hypothesis-fatigue." Perhaps someone threw away some seeds he had gathered for food and found, to his amazement, that they grew in profusion. The difficulty is that lucky accidents would lead to domestication only if man were prepared to take advantage of them. As the British scholar A. M. Hocart has noted, "Man has been acquainted with electricity for at least two thousand years, but he never saw how he could make use of it till the last century. His accidents served him no purpose until he had been prepared to profit from them by centuries of physical investigation." [26]

The accident thesis is an "inevitability" thesis. Raymond A. Dart, for example, argues for multiple origins of domestication wherever suitable conditions existed, which he defines as the presence of a river and lake-bank fishing society.[27] Frederick E. Zeuner looks on domestication as the inevitable outgrowth of the relationship between man and animals, given their respective biological and psychological characteristics. Both men and animals are domesticating creatures in this view. Men domesticate other men and it is called slavery. Examples of animal symbiosis are numerous, ranging from perfect symbioses where both partners benefit without suffering, through unequal partnerships where one species benefits and the other does not suffer, to social parasitism. Zeuner believes that a certain level of social life must have been reached by both domesticator and domesticated before domestication became possible. Animals that live in herds or packs are more willing to enter into social relations with members of other species. In Zeuner's view there are various ways in which social relations between men and animals can be established, many of which have their parallel in the animal world—for example, through the keeping of pets and scavenging. The domestication of reindeer, presumed to have been the accomplishment of hunters, is to Zeuner an example of pure social parasitism.[28]

While there is unquestionably much truth in Zeuner's analysis it leaves basic questions unanswered. Why have some herd animals been domesticated and not others? Why have some men domesticated and not others, although often those who did not domesticate lived on the same level of social organization as those who did and were in contact with domesticable herd animals? Why has considerable effort gone into the attempt to domesticate nonherd animals whose psychology makes them peculiarly difficult to domesticate, e.g., snake, cat, and fowl? Zeuner does outline the basic human and animal characteristics that made do-

[26] *Social Origins* (London: Watts and Co., 1954), p. 130.
[27] *Africa's Place in the Emergence of Civilization* (Johannesburg, South Africa: South African Broadcasting Corporation, n.d.), pp. 33–46.
[28] Frederick E. Zeuner, *A History of Domesticated Animals* (New York and Evanston, Ill.: Harper & Row, Publishers, 1963), pp. 46–48.

mestication possible, but his theory does not explain why domestication occurred at particular times and places.

What has happened is that the conditions under which domestication can take place have been converted into the cause or explanation for domestication. Oakes Ames, whose *Economic Annuals and Human Cultures* is a classic in the literature of plant domestication, makes the obvious point that man used the flora extensively for such a long time that knowledge of which plants are useful and which are not grew steadily. Eventually the transition to cultivation and domestication occurred.[29] In other words, the domestication of plants was bound to happen. And so we are back to a view that is subject to many of the same strictures we applied to Zeuner's argument. Students of New World domestication have been especially attracted to this notion of domestication arising as a response to environmental or inherent socio-cultural forces: by and large, archaeologists and anthropologists who have concerned themselves with the rise of New World agriculture have assumed independent New World domestication.

Explanations of domestication in terms of the conditions that make it possible, for all their apparent reasonableness, have implications that are not immediately obvious. If the facts were completely other than what they are, the "conditions" could explain them equally well. For example, it might be argued that man's condition as the only predatory carnivorous species among the primates, able to digest meat in the raw state but unable to digest most vegetables before he learned to master fire, would necessarily lead him to domesticate animals before turning to plants. The only trouble with this explanation is that it seems clear that man domesticated plants before animals. And had man not domesticated plants at all, it could have been argued that time and energy requirements for the preparation of vegetable food for human consumption were so high that man never found it worthwhile to domesticate plants.

While the stage theory is finished and the evolutionary theory is gone at least for the foreseeable future, the geographic hypotheses concerning domestication seem to have come full circle. Formerly it was widely believed that the Near East was the hearth of domestication, and that the three stage ladder started there. The early great civilizations grew up in its river valleys, and it seemed reasonable to believe that the invention of domestication had formed the basis for the civilization of its creative peoples. But then came impressive archaeological finds of the European Mesolithic and Neolithic, and a number of scholars believed that Europe must have been the earliest center of domestication. Subsequently many scholars hypothesized a tropical origin for domestication, largely because the ethnological work being done in Africa and Asia as well as in the New World tropics, focussed attention on this area. The predominant view at present is that the Old World hearth is, as Sauer

---

[29] Oakes Ames, *Economic Annuals and Human Cultures* (Cambridge, Mass.: Botanical Museum of Harvard University, 1939), p. 127.

urges, in southern Asia in a closely defined area within the tropic realm. But there are signs of a change from this view also, and the Near East is again emerging as the most probable hearth of domestication. Obviously archaeological evidence, which clearly points to a Near Eastern hearth, is not sufficient. But more and more evidence of quite different sorts, as subsequent chapters in this book will attempt to show, supports a Near Eastern hearth for domestication.

CHAPTER 2 *interpreting the archaeological record*

The archaeological record provides a logical beginning for any attempt to answer the where, when, and why of the domestication of plants and animals. Yet despite the dramatic achievements of twentieth century archaeology, grave doubts about the ability of archaeology to answer questions of origins remain. For although archaeology can presumably establish sequences and even cast light on the process of change, it cannot discover where the first crucial step was taken. As Carleton S. Coon has said, "It is too much to ask of the law of chance that the first archaeologist who digs a hole in the ground should locate the place where the invention took place." [1] Many holes have been dug from the Near East to the rainy tropics and elsewhere, but to hope that the earliest domestic remains of plants or animals or even implements of agriculture have been found in them is illusory. It seems even more illusory to expect archaeology to answer the "why" of domestication. For even assuming, by some leap of faith, that the earliest domestic grain found were in fact the first domestic grain, archaeology could say very little about the *reason* for the transition from the incredibly old food collecting and harvesting or specialized hunting economies to systematic care and control of crops and animals.

For the study of the origin of domestication, the conventional indices for determining the existence of agriculture in an archaeologically discovered culture are problematical. In the first place, cultivation of crops does not imply that the crops are domestic. Most crops must have been cultivated long before the characteristics of domestication appeared. There are plants that are cultivated, or at least protected and occasionally and intentionally propagated, and at the same time harvested where they

[1] *The Seven Caves* (New York: Alfred A. Knopf, Inc., 1957), p. 151.

occur as wild stands. This is particularly true of many berries, mushrooms, cactuses, herbs, and many tree crops in which few or none of the features generally associated with domestication have appeared even today (e.g., raspberry, wild rice, sea buckthorn). Strictly speaking, cultivation and domestication, although very closely connected, are not synonymous. The distinguishing characteristics of domestication have often appeared without cultivation and without any sign of protective human interest in the plants concerned. Gigantism, accelerated fruiting, and other desirable properties of domestic plants often occur in plants growing in camp refuse, for example. This fact has given rise to some theorizing about a "dump heap" origin of domestication.[2]

Further limiting the evaluation of archaeological finds is the undismissible possibility that the first plants domesticated may not be our current crop plants. Many plants formerly cultivated have been given up entirely (e.g., goosefoot, *Chenopodium;* skirret, rampion) or given up in areas where they were formerly important (e.g., the water chestnut in Europe). Perhaps plants considered utterly useless today, such as most of our field weeds, were the first to be systematically propagated, and perhaps many of our present-day domesticates were originally weeds in the stands of formerly noble plants. It has been suggested that some "flax mimics," perhaps *Camelina sativa* subsp. *linicola,* were unsuspecting host to flax which ultimately degraded its host to the position of a weed in flax fields.[3]

Ethnologists and cultural geographers have long been impressed by the extraordinary intervention in natural vegetation effected by contemporary primitives. They have shown in detailed studies that fires set by man have not only changed the floristic composition of plant associations but also have established major plant formations such as savannas. Thus while the plant geographer A. F. W. Schimper still believed that because of their zonal distribution savannas were natural formations, the fire origin of most grasslands is today widely accepted. Many scholars have drawn the obvious implication that primitive and prehistoric burning produced a man-made or *anthropogene* vegetation. Indeed, many genetic changes affecting fire resistance, life span, seed production, etc. can be derived directly from human activity.[4] Widespread burning for

[2] E.g. Edgar Anderson, *Plants, Man and Life* (Berkeley and Los Angeles: University of California Press, 1967), 136–50.

[3] On flax see Bruce Wallace and Adrian M. Srb, *Adaptation,* 2nd ed. (Englewood Cliffs, N.J.: Prentice-Hall, Inc., 1964), pp. 46–48. On flax and its mimics and on the general problem of weeds, domesticates, and cultivation see A. Thellung. "Kulturpflanzen-Eigenschaften bei Unkräutern," in *Festschrift Carl Schröter,* ed. H. Brockmann-Jerosch, Veröffentlichungen des Geobotanischen Institutes Rübel in Zürich, 3. Heft (Zürich: Kommissionsverlag von Rascher & Co., 1925), 745–62.

[4] A. F. W. Schimper, *Plant Geography upon a Physiological Basis,* trans. W. R. Fisher, rev. and ed. by Percy Groom and Isaac Bayley Balfour (Oxford: Clarendon Press, 1903), pp. 260–83, 363–76. On the effect of fire see Carl O. Sauer, *Agricultural Origins,* pp. 8–18, and "Fire and Early Man," in *Land and Life, A Selection from the Writings of Carl Ortwin Sauer,* ed. John Leighly (Berkeley and Los Angeles: University of California Press, 1963), pp. 288–99. Other essays pertinent to the effects of burning on vegetation are included in this volume. See also Omer C. Stewart, "Fire as the First Great Force Employed by Man" in *Man's Role in Chang-*

the purposes of food production is not confined to farming societies. For example Auen Bushmen burn to increase yields of tubers. Other Bushmen in Botswana and Rhodesia set fires to bring up fresh grass in order to attract into the burned over areas the animals they hunt.

Anthropogene vegetation, logically at least, is very close to plant domesticates. In both cases the changes in the plants have been called forth by human activity, and both anthropogene vegetation and domesticated plants depend on continuing human activity. Many scholars have therefore related domestication to that human presence in an area which affects vegetation. Thus if the Auen burn in order to increase the size and yield of gathered plants, or if tribes on the Chukchi peninsula collect plants growing in the refuse accumulation near their tents, they are in the view of some interpreters ". . . so to speak, unintentional plant breeders." One "volunteer" plant gathered by the Chukchi "a cineraria, a composite plant, deserves particular mention, for it is found only around the tents, where it contributes its annual share to the support of the Chukchi."[5] It appears then as self-evident to many students that the transition from "foraging," "scavenging," "food gathering," "unspecialized hunting and collecting," or "specialized hunting and food collecting," to domestication involves a state of "semi-" or "quasi-domestication" of plants (and/or animals) which is probably of such antiquity that it lies far beyond the agricultural societies that can be recovered by archaeology.[6]

While social, cultural, and ecological considerations are of the greatest importance in the study of domestication, it is essential to realize the limitations of such approaches when archaeological and sometimes physical anthropological verification is absent. Inevitably theses based on the life of present-day primitive societies assume that such societies reflect the economic and social conditions of the Paleolithic or Mesolithic and ignore the possibility of cultural regression and breakdown. Australian natives who are both skilled hunters and food collectors are thought by some to derive from a remote farming substratum, for there is evidence that many of the tribes formerly had a system of matrilineal descent and

---

ing the Face of the Earth, ed. William L. Thomas Jr. (Chicago: The University of Chicago Press, 1956), pp. 115–33. For further references see Henry J. Oosting, *The Study of Plant Communities* (San Francisco, Calif.: W. H. Freeman and Company, 1950), pp. 204–5, 255–56, 282, and Josef Schmithüsen, *Allgemeine Vegetationsgeographie* (Berlin: Walter de Gruyter & Co., 1961), pp. 133–34.

[5] Quoted after A. Maurizio by Franz Schwanitz, *The Origin of Cultivated Plants,* trans. Gerd von Wahlert (Cambridge, Mass.: Harvard University Press, 1966), p. 12.

[6] Cf. Oakes Ames, *Economic Annuals and Human Cultures,* pp. 139–40. Terms such as "specialized hunters," "plow complex," etc. have been used by many cultural historians, geographers, and ethnologists either as specific designation for a hypothetically reconstructed culture within specified geographic boundaries or simply as designations for the characteristic aspects of an economy whether ancient or not. The notion of "cultural complex," "culture area," as well as the idea of Kulturkreis are critically examined by Paul Leser in "Plow Complex, Culture Change and Cultural Stability," in *Selected Papers of the Fifth International Congress of Anthropological and Ethnological Sciences, Philadelphia, September 1–9, 1956,* ed. Anthony F. C. Wallace (Philadelphia, Pa.: University of Pennsylvania Press, 1960), pp. 292–97, and the same author's "Zur Geschichte des Wortes Kulturkreis," *Anthropos,* 58 (1963), 1–36.

matrilocal marriage, which are widespread among contemporary primitive cultivators.[7] Even where modern food collectors show no symptoms of relationship to other types of economy, they have coexisted with and been influenced by different cultures, so that their ways of thinking and acting may differ radically from those of prehistoric man. Pygmy food collectors of the eastern Congo have many traits in common with Bantu farmers and, as their bow and other sophisticated hunting gear attests, with some Upper Paleolithic and Mesolithic specialized hunting cultures (e.g., the Aterian of North Africa, the Magdelenian-Hamburgian of west-central Europe). Finally, as far as is known, none of the collecting and hunting cultures cited as examples of "quasi-domesticators" have in fact until now made the transition to a herding or farming way of life.

In turning to the archaeological record, there are two practical problems: 1. Are the animal or plant remains in a given archaeological find domestic? 2. In the absence of characteristic features of domestication, is there cultural evidence pointing to cultivation or animal husbandry, presumably in this case of species that have not yet acquired domestic traits? If there is such cultural evidence, the culture may be considered a domesticating culture in spite of the absence of morphological characteristics of domestic remains. The definition of domestication thus involves both cultural and morphological characteristics which need not both be present for the finds to be judged domestic.

Since animal remains are generally better preserved and on the whole older than domestic plant remains (for all that plant domestication is in all likelihood much older than animal domestication), we shall consider the domestic traits of animals first. Animals are fully domesticated when the following characteristics are found:

1. The animal is valued and there are clear purposes for which it is kept.
2. The animal's breeding is subject to human control.
3. The animal's survival depends, whether voluntarily or not, upon man.
4. The animal's behavior (i. e. psychology) is changed in domestication.
5. Morphological characteristics have appeared in the individuals of the domestic species which occur rarely if at all in the wild.

There are animals who meet some of these criteria but not others. The elephant's breeding, for example, is not controlled by man. Rats and houseflies are generally not valued and are not kept for human purposes. Yet so-called domestic varieties exist in antipathetic symbiosis with man. Individuals of many animal species may be tamed, but taming does not constitute domestication unless the species itself meets all of the five criteria enumerated above. Animals that meet some but not all of these criteria can at most be called semidomestic. Archaeologically it is possible to find evidence that a certain measure of control was exercised (e.g., enclosures were built), but it is impossible to be certain if the animals

[7] Karl J. Narr, "Das Höhere Jägertum: Jüngere Jagd- und Sammelstufe" in Historia Mundi, ed. Fritz Valjavec, Erster Band: Frühe Menschheit (Bern: Francke Verlag, 1952), p. 504.

so enclosed were fully domestic. It therefore becomes necessary to rely almost wholly on morphological characteristics in evaluating archaeological evidence.

In the last sixty years great strides have been made in the comparative study of domestic and wild individuals of a species, and many changes that are the result of domestication, including changes in the soft parts of the body reflected in skeletal remains, have been clearly established. Theoretically it should now be possible to distinguish between wild and domestic animals in early finds and representational art. In cattle, for example, a foreshortened and widened skull, decrease in the dimensions of eye and ear openings, shortness of backbone, decrease in size—in short, overall infantilism—distinguishes domestic from wild varieties. It is remarkable that many changes are common to animals of very different species: curly hair instead of straight; retention of baby hair; pied instead of monocolored coats; reduction in differences between male and female; variability in size between different breed groups leading to a pronounced contrast between giants and dwarfs; extremely one-sided development of certain characteristics, such as milk production; and sometimes pathological alterations, such as the short-leggedness of Dexter cattle, in which the gene responsible is lethal when homozygous. Some of the changes in the soft parts are reflected in skeletal remains. Muscular development or atrophy and changes in brain volume due to environmental modifications, such as differences in food supplied by man or the specialized physiological performance required of domestic animals, mark the skeleton and lead to the development of characteristic crests or ridges.[8]

But the usefulness of these criteria in the examination of skeletal remains from the dawn of domestication is limited: cultural domestication antedated any impact upon the osteological parts of the animal. The difficulty is aggravated by the fact that bones to which such diagnostic criteria may be applied are often missing or have been discarded in archaeological finds. Nor can we exclude the possibility that the changes occurred in wild mutants, for almost all the changes that occur in domestication are known to occur, although rarely to be sure, in wild individuals. Thus for all the progress that has been made in identifying characteristics that develop in domestication, these criteria are insufficient for determining whether domestication had in fact occurred in the earliest sites where predomestic animals are found. Indeed, as Hans Epstein, a leading student of African domestic animals, points out, the study of anatomic characteristics has been inadequate even for determining the racial history of longhorn cattle.[9]

[8] For citations to the relevant literature on changes attributable to domestication see Erich Isaac, "On the Domestication of Cattle," *Science* 137, No. 3525 (July 1962), 195–204, particularly p. 204, note 31. See also M. L. Ryder, "The Exploitation of Animals by Man," *Advancement of Science*, 23, No. 107 (May 1966), 9–18, especially pp. 10–11.

[9] Die Unbrauchbarkeit einiger anatomischer Merkmale für die Rassengeschichte europäischer Langhornrinder," *Zeitschrift für Tierzüchtung und Züchtungsbiologie*, 71, No. 1 (1958), 59–68. On the unfortunate treatment of animal remains in many excavations see the lament of Raymond E. Chaplin, "Animals in Archaeology," *Antiquity* XXXIX (1965), 204–11, especially p. 204.

The presence of tools that are clearly elements of animal husbandry extends the usefulness and the range of the archaeological record. But the purpose to which many early tools were put is uncertain; and where the association with domestication is clear, as for example goads and animal harnessed plows, the tool is later than domestication of the animal. In the absence of clearcut characteristics of domestication and of artifacts whose purpose can be seen clearly, there has been since the turn of the century emphasis on the use of statistics to indicate cultural control over animal populations. The bone assemblage of a site may show a change from a great range of animals killed in the oldest periods to a prevalence of prodomestic animals, that is, animals which are known to have become domesticated at a still later period. Subsequently further shifts may occur in the proportions of bones of male and female, old and young prodomestic animals. A preponderance of adult bones may indicate domestication. Presumably adults had to be killed in order to capture their young whom they tried to protect. Similarly, a preponderance of male bones has been taken to indicate incipient domestication, for the domesticator saved the young males and females.[10] Conversely, shifts to a preponderance of submature animals have also been taken to show a cultural control of the herds.[11]

For plants it is possible to specify in principle a number of conditions that must be met before the plant can be considered a domesticate. Such conditions parallel the requirements enumerated for domestic animals. (There are differences—obviously it makes no sense to speak of a change in "plant psychology.") If there are ambiguities in the archaeological evidence concerning animals, allowing interpretations of bone finds to swerve wildly between "domesticated," "semidomestic," and "wild and hunted," the ambiguities are much greater in the case of plants. Useful plant material is in any case difficult to recover in the oldest archaeological remains. Furthermore, while morphological and cytological changes have occurred in many domestic plants, in others they have not.

Strictly speaking, only plants in which morphological changes have occurred and which are valued for their economic contribution should be called "domestic." The wider category "cultivated" would include both domesticated plants and cultivated "wild plants" in which there has been no morphological change. There are also plants which are neither cultivated nor domesticated, but are nonetheless extensively used, as for example, the baobab tree of the African savanna, the maté plant of South America, and the coconut palms of the Indian Ocean region. While

[10] S. Bökönyi, "Archeological Problems and Methods of Recognizing Animal Domestication." In Peter J. Ucko and G. W. Dimbleby, eds., *The Domestication and Exploitation of Plants and Animals* (Chicago: Aldine Publishing Company, 1969), pp. 219–29.

[11] Robert H. Dyson, Jr., "Archeology and Domestication of Animals in the Old World." *American Anthropologist*, 55, No. 5, Pt. I (1953), 662. Concentrations of submature animals occur also. See Charles A. Reed, "Animal Domestication in the Prehistoric Near East," *Science*, 130, No. 3389 (December 1959), 1635, for the Egyptian site of Toukh.

the use of such trees and plants, and the protective attitude often adopted toward them, has played a role in some theories of domestication and is of interest to ethnobotanists, it is of marginal concern here. Given cultural and economic incentives, many of these plants could be cultivated, as indeed some are in plantation or orchard systems. Many medicinal herbs and relish plants are cultivated, and ultimately domesticated varieties may be developed.[12] Weeds such as flax mimics may be formerly domestic plants that are no longer deliberately cultivated. Weeds may continue their association with cultivated crops, in which case they may be recovered with domestic plants in a deposit, or they may have migrated beyond the limit of cultivation, or again they may have remained in occupation of an area once cultivated and now abandoned. Important as weeds could be in historical geographic reconstructions of land use and in studies of dispersion of domesticates, little has been done with them.

Morphological or cytological changes that distinguish domestic plants from their wild ancestors have been summarized by Franz Schwanitz. Domestic plants increase in size and robustness, often because of polyploidy (the doubling or multiplication of chromosome numbers). They decrease in fruit producing ability, but this decrease is accompanied by an increase in the size of individual fruits. Domestication leads to an increased number of varieties. In addition to the general gigantism of the domestic plant, there is a disproportionate (allometric) growth of nonharvested parts. The fruit of the domestic red pepper, for example, is approximately 500 times greater in weight than that of the wild variety. The cultivated plant loses wholly or in part its ability to disseminate itself, relying on man for propagation. Loss of bitterness and toxicity which protects wild plants from their enemies is another hallmark of the domestic plant. Insufficiently watered, the bitter cucumber serves as a reminder of its wild ancestor. Loss or reduction of mechanical means of protection, like glumes or awns in cereals, is another feature of domestic plants. In addition, while the wild species have a tendency to produce fruit over a long period of time, the fruit of domestic species ripen simultaneously.[13] There are domesticated plants, such as some citrons, which bear fruit through the seasons, but in this case deferred ripening may have been intentionally preserved by man. The life span of the domestic varieties is generally shorter than that of the wild. Wild rye, for example, is a perennial, but cultivated winter rye an annual.

Actual plant finds in the archaeological record are few and depend upon the accident of preservation. Finds consist of carbonized materials, imprints of grain and seed in pottery and sun dried adobe, lignified material such as food deposited in tombs, and silica skeletons of chaff or

---

[12] Cf. Claude Lévi-Strauss, "The Use of Wild Plants in Tropical South America," in *Handbook of South American Indians,* Vol. 6 Smithsonian Institution, Bureau of American Ethnology Bulletin 143, ed. Julian H. Steward (Washington, D.C.: Government Printing Office, 1950), pp. 465–86. On the Indian Ocean coconut see Jonathan D. Sauer, *Plants and Man on the Seychelles Coast* (Madison, Milwaukee, and London: The University of Wisconsin Press, 1967), pp. 24–50.

[13] Schwanitz, *Cultivated Plants,* pp. 14–63.

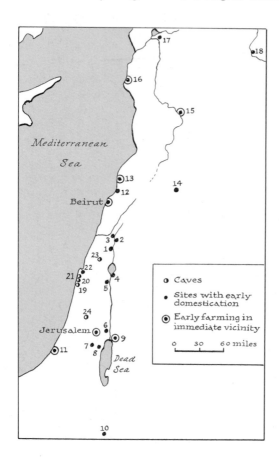

Figure 1. *Selected Early Sites in the Near East (I). 1, Einan; 2, Hagoshrim; 3, Maayan Baruch; 4, Shaar Hagolan; 5, Hamadia; 6, Jericho; 7, Erq el-Ahmar; 8, El Khiam; 9, Ghassul; 10, Beida; 11, Askalon; 12, Ras el-Kelb; 13, Byblos; 14, Yabrud; 15, Hama; 16, Ras Shamra; 17, Judeideh; 18, Mureybat; 19, Kebara; 20, Tabun; 21, El-Wad; 22, Nahal Oren; 23, Qafza; 24, Shukbah.*

stomach contents of prehistoric corpses preserved in peat bogs. If a cardinal sin of archaeologists was throwing out bones without investigation, the sin was not even consciously committed in the case of plants whose remains were not even seen.

The rarity of plant finds and the difficulty of determining by macroscopic and microscopic analysis whether the plant was domesticated has meant that the presence of tools associated with plant cultivation has been crucial in the attempt to pinpoint incipient domestication of plants. While the statistical approach gives some idea of when prodomestic animals appear, that approach is difficult if not impossible for plants. Helbaek conjectures, ". . . probably some kind of nursing and protection of wild food plants came first, without either actual specialized implements or sedentary life, while only eventually did the advantages of tillage and of permanent residence dawn upon the earliest farmers." [14] The actual turn-

---

[14] Hans Helbaek, "The Palaeoethnobotany of the Near East and Europe," in Braidwood and Howe, *Prehistoric Investigations*, pp. 99–119. Quote is on p. 100. Henceforth quoted as *Palaeoethnobotany*.

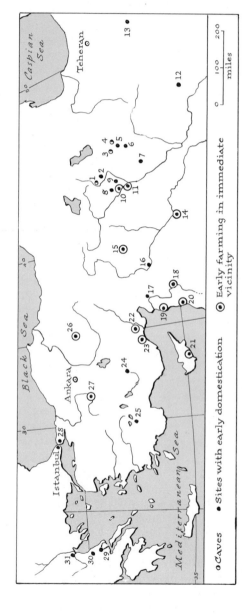

Figure 2. *Selected Early Sites in the Near East (II).* 1, *Shanidar;* 2, *Zawi Shemi;* 3, *Zarzi;* 4, *Pale-gawra;* 5, *Karim Shahir;* 6, *Jarmo;* 7, *Matarra;* 8, *Tepe Gawra;* 9, *Arpachiya;* 10, *Nineveh;* 11, *Hassuna;* 12, *Ali Kosh;* 13, *Sialk;* 14, *Mari;* 15, *Tell Halaf;* 16, *Mureybat;* 17, *Judeideh;* 18, *Hama;* 19, *Ras Shamra;* 20, *Tabbat el-Hammam;* 21, *Khirokitia;* 22, *Tarsus;* 23, *Mersin;* 24, *Catal Hüyük;* 25, *Hacilar;* 26, *Aliçar;* 27, *Yazirhüyük;* 28, *Fikirtepe;* 29, *Sesklo;* 30, *Argissa Magula;* 31, *Nea Nikomedeia.*

25

ing point from the food collecting to the food producing stage in human economy, in Helbaek's view, can never be demonstrated fully either botanically or culturally. In addition, it is impossible to determine whether tools we find were used for the care or harvest of wild or domestic plants.

Customarily discussions of domestication begin with the Neolithic, defined economically by the emergence of food production to replace the older hunting and gathering ways of life. Often a Mesolithic period intermediate between Neolithic and Upper Paleolithic cultures is identified and attributed with "incipient" domestication of plants and animals. However, there is reason to believe that the roots of domestication lie in the Upper Paleolithic, roughly 35,000 to 14,000 years ago, when the great Eurasian region, in which cultures spatially distant from each other had been similar for long periods of time, gave way to cultures of much more limited range which evolved more rapidly. The oldest known Upper Palaeolithic sites in which incipient domesticators may have lived are Near Eastern. Moreover, the Mesolithic and Neolithic sites in that area are considerably older than other sites in which similar economies have been found.

Some of the chief Upper Paleolithic sites in the Near East are Erq el-Ahmar, the caves of El-Wad and Kebara on Mount Carmel, Atlit in the coastal plain of Israel near Mount Carmel, the caves of Jebel Qafza and Emire in Galilee, the rock shelters of Yabrud in the Syrian Anti-Lebanon, and Ksar Aqil on the Lebanese coast. Other important sites are Shanidar in Iraqi Kurdistan and Belt and Hotu Caves in the southern Caspian region. Archaeology reveals that these cave sites or rock shelters were occupied for long periods by the same or similar human groups. The record is not clear as to whether the people were fully sedentary or came and went repeatedly over undetermined time intervals, the latter interpretation being customary. Nonetheless, many of the Middle and Upper Paleolithic sites have characteristics that point to a more than intermittent occupation. Caves in close proximity to each other were occupied simultaneously by peoples of very different cultures. If they were not occupied continuously, it is hard to explain why neighboring groups of different culture did not take over the site. It is possible that these people were not fully sedentary, but that property in habitation sites, even in the absence of the "owner," was respected for religio-magical reasons. Even in the Mousterian, which precedes the Upper Palaeolithic, property in place seems to have extended to the dead. Certain caves such as Skhul on Mount Carmel served as burial grounds for many generations, while nearby Tabun cave, simultaneously occupied for a long time, had hardly any burials.[15] In addition to caves and rock shelters, the Upper Paleolithic in Palestine (as indeed in central Europe and the Russian grasslands) shows evidence for the occupation of open-air camp sites. The Near Eastern "camp sites" are particularly well preserved, not having

---

[15] Emmanuel Anati, *Palestine Before the Hebrews* (New York: Alfred A. Knopf, Inc., 1963), pp. 100–101, 128–29.

been disturbed by subsequent agriculture or vegetation. The absence of stratification and the thinness of material suggests that most camps were occupied on a temporary basis.

The very idea of property may have been important for the origin of domestication. For if property rights in a place were so strong that they were protected even in the absence of the owners, similar notions may have been applied to particular stands of plants held to belong to individuals or small groups. Indeed the protective attitude toward plants often assumed to characterize incipient domestication may itself be based on a primary experience of property. Among students of modern primitive food collectors, Father Wilhelm Schmidt interprets the tacitly observed collecting-range boundaries as the beginning of land ownership. Others interpret these more narrowly as pertaining to the usufruct of certain plants, as reward for the finder's effort or producer's labor. These divergent interpretations are not mutually exclusive, and a protective attitude toward one's own plants may have been conducive to attempting their propagation and multiplication—on one's own land.[16]

A settlement pattern combining permanent cave dwellings with temporary camps then seems to have been characteristic of the Upper Palaeolithic in the Levant and perhaps also of western Iran. This suggests that there was a migratory, probably seasonal range for at least some population groups. While there is no archaeological evidence for domestication or even for cultivation of plants in the Upper Palaeolithic of the Near East, the conditions under which cultivation was possible were met.

Sauer believes the necessary conditions for the primary domestications, were met in southeastern Asia.[17] However, it may be that the Near East of the Upper Paleolithic met certain of Sauer's requirements. The climate at that time, in the last pluvial, was not the climate of today. During Middle Paleolithic times the climate was similar to that of present-day monsoon Asia and the fauna, which included elephants and hippopotami, indicate tropical savannas. In the Upper Paleolithic, after the major interstadial of the last pluvial, the climate was much drier, but still wetter than it is today. The fauna had become modern.[18] Both the climate and the sedentary tendencies of the people, who at most seem to have migrated seasonally over short distances, provide grounds for hypothesizing some form of plant cultivation had its inception here. The seasonal migration pattern lends credence to this, since the open air sites

---

[16] Wilhelm Schmidt, *Das Eigentum auf den Ältesten Stufen der Menschheit* (Münster i. W.: Aschendorff, 1937), I, 215. Josef Haekel, "Zum Problem des Mutterrechts," *Paideuma* 5 (1953), 298–322. W. Nippold, *Die Anfänge des Eigentums bei den Naturvölkern und die Enstehung des Privateigentums* ('S-Gravenhage, Holland: Mouton & Co., 1954), 13–16, 48–52. Erich Isaac, "God's Acre, Property in Land: a Sacred Origin?" *Landscape*, 14, No. 2 (Winter 1964–1965), 28–32.

[17] *Agricultural Origins*, 22–24.

[18] These conclusions are offered with some diffidence. On the weaknesses of climatic reconstructions in this region see Wright, "Climate and Prehistoric Man," pp. 84, 87–88.

were in grasslands and the purpose of the seasonal visits may have included the harvesting of wild grasses.[19]

The transition to the Mesolithic seems to have come about earlier in the Near East than anywhere else, 14,000 to 12,000 years ago or several thousand years earlier than in Europe or Africa. Whether the conjecture that plant cultivation was practiced in the Upper Paleolithic is correct or not, it is highly probable that in the Mesolithic domestication of plants and animals occurred. Specifically, the Natufian culture, first discovered by Dorothy Garrod at Shukbah cave in Wadi en-Natuf in central Palestine, is the first of the Mesolithic cultures to offer convincing proof for the beginning of cultivation.[20]

In contrast to Upper Paleolithic cave and open air sites, whose location shows no clear topographic pattern, Natufian settlements are along lakes, marshes, rivers, brooks, and seashore. It is interesting that the Mesolithic both of Europe and of tropical Asia included hunting and fishing cultures. The European Mesolithic, despite its use of elaborate hunting and fishing gear, appears impoverished compared to the preceding Magdalenian with its remarkable artistic achievements. Only in the Near East did the Mesolithic produce a tremendous burst of energy manifesting itself not only in the transition to agriculture but in an art unparalleled in other Mesolithic cultures. Most of the surviving Mesolithic art consists of sculptures of man or animal decorating sickle hafts or mortar pestles.

The importance of cereal foods in the Mesolithic is shown by the large number of sickles and mortars found in Natufian sites. The in-

[19] Sauer, *Agricultural Origins,* pp. 24, 27, points out that the use of poisonous plant extracts for fishing is most elaborated in Southeast Asia and he believes that the interest in this method of fishing may have been a milestone to agriculture. But the ability to extract plant poisons must have long preceded Mesolithic fishing cultures. Even gatherers must have learned to rid plants of their poisons, or else the range of available foods would have been extremely narrow. Schwanitz observes: "Of a rough total of 51 plants gathered for food by the aborigines of Australia, only 36 can be eaten raw. None of them is pleasant tasting or of great nutritive value. Nine plants in a second group have roots that are edible if baked. The remaining six plants, which provide the most important vegetable food for the aborigines, are all toxic in the raw state." (*Cultivated Plants,* 39) Toxicity is not limited to tropical or subtropical plants or to tubers. Millets must be scalded and the water discarded before cooking in order to produce a savory dish. In short, even though Sauer may be correct in his assessment of the criteria which a domesticating society must meet, these criteria may have been met in other societies, including those of the Upper Paleolithic Near East. Nor is there any necessary coincidence between a center of elaboration of a trait and its center of origin. Cf. Robert F. Heizer, "It is probable that fish stupefying was not invented by coastal people but probably by interior peoples . . . to whom fish was probably a subsidiary item of the dietary." *Aboriginal Fish Poisons* (Washington, D.C.: Smithsonian Institution, Bureau of American Ethnology Anthropological Papers, No. 38, 1953), 236. The poisonous nature of early domesticates is stressed by H. Brockmann-Jerosch, "Die Kulturpflanzen, ein Kulturelement der Menschheit," in *Festschrift Carl Schröter,* Veröffentlichungen des Geobotanischen Institutes Rübel in Zürich, ed. H. Brockmann-Jerosch (Kommissionsverlag von Rascher & Co., 1925), pp. 793–810. Especially p. 798.

[20] D. A. E. Garrod, "Excavations at the Cave of Shukbah, Palestine, 1928," *Proceedings of the Prehistoric Society,* n. s. VIII (1942), 1–20. On the relation of the Natufian to other Mesolithic and Neolithic industries in the region see M. W. Prausnitz, "A Study in Terminology: The Kebaran, the Natufian and the Tahunian," *Israel Exploration Journal,* 16, No. 4 (1966), 220–30.

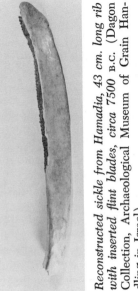

Reconstructed sickle from Hamadia, 43 cm. long rib with inserted flint blades, circa 7500 B.C. (Dagon Collection, Archaeological Museum of Grain Handling in Israel)

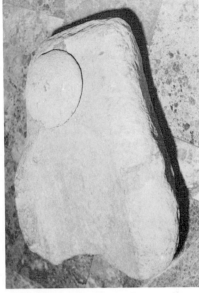

Prepottery quern and grinding stone from Nahal Oren. The limestone quern is 47 cm. long, 19 cm. high and maximum width is 36 cm. Dates from circa 8000 B.C. (Dagon Collection, Archaeological Museum of Grain Handling in Israel)

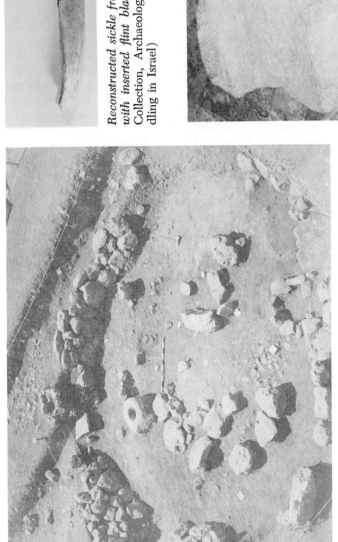

House and mortar as excavated at Einan by Jean Perrot. (J. Perrot, Mission Archéologique Française en Israel)

creasing importance of cereals is apparent in the fact that early Natufian microliths were more carefully worked than later ones used in composite sickles: The spread of these harvesting implements diminished the importance of carefully working a single tooth. In the cave of El Wad on Mt. Carmel fragmentary remains of thirteen sickles were found. As the prehistorian Emmanuel Anati has noted, "Whatever was collected and ground and mashed by these tools must have been plentiful enough to justify the abundance of these tools."[21]

Knowledge about the Natufian was expanded sharply with the sensational finds of Jean Perrot at Einan beginning in 1955 and those of Kathleen Kenyon at Jericho since 1952. At Einan, an open-air Natufian settlement was uncovered that was similar in plan to the small open-air camp sites of the Upper Paleolithic, but extending for more than a thousand square meters. Circular plastered storage basins and pits, and very heavy stone basins were also found.[22] The heavy basins and the large number of burials, the latter indicating a cemetery tradition, argue against a nomadic way of life. Furthermore, the settlement was rebuilt a number of times, providing strong evidence for prolonged sedentary life there. At Jericho Kathleen Kenyon found evidence of what she terms a proto-Neolithic settlement, dating at least to the ninth millennium B.C. The earliest level contains a succession of the floor levels of huts of clay, wood, and skins. The earliest levels are followed by evidence of an actual town with buildings which were round or ovate, slanting toward beehive shaped roofs: a radio carbon date of ca. 7800 B.C. is given for this level. Around 7,000 B.C. this prepottery town was surrounded by an enormous defensive wall of massive stones, some weighing several tons, and surrounded with a rock-cut moat. Even more remarkable was the discovery of a great round stone tower and evidence that there may have been more. Anati believes that these towers were part of the fortifications. Other prepottery or aceramic Neolithic villages have been found in Nahal Oren on Mount Carmel, Khirokitia on Cyprus, Jarmo in Iraqi Kurdistan, Haçilar in Anatolia, and more problematical, Beida in Transjordan, Mersin in Cilicia and Byblos in Lebanon. Jericho, however, differs sharply from these farming villages, for it is the only example of a fortified town before the fourth millennium B.C.[23]

There has been argument over whether the "Natufians" were actually domesticators of plants. The mortars, sickle hafts, and pestles show that grains were harvested, but they may have come from wild stands. It has been argued that cereals could not have been harvested for their fruit, for wild grasses have a very fragile rachis that would shatter if the stalks were cut with a sickle. Thus sickles, it is said, may only imply that wild grass was harvested to provide thatch. But even in the small Natufian village communities, the size of the mortars, which were very large and

[21] Anati, *Palestine*, p. 149.
[22] J. Perrot, "Excavations at 'Eynan ('Ein Mallaha)," *Israel Exploration Journal*, 10, No. 1 (1960), 14–22. On Jericho see K. M. Kenyon, *Digging Up Jericho* (New York: Frederick A. Praeger, Inc., 1957) and *Archaeology in the Holy Land* (New York: Frederick A. Praeger, Inc., 1960), pp. 36–100.
[23] Anati, *Palestine*, pp. 242–43.

heavy and could not have been transported by a nomadic group, argue against this hypothesis. The discoveries of Einan and Jericho destroy it altogether, for Einan is Natufian throughout and Jericho is Natufian in its earlier layers, including the period of its earliest fortifications. Since prepottery Jericho is estimated to have supported a population of at least 2000, agriculture must have been practiced. It goes without saying that the cereals that were harvested must have been well on the way toward the development of a tough rachis, which puts their incipient domestication at least as far back as the early ninth millennium B.C.

The chief question posed by Jericho is why it should be unique. There is nothing in Jericho's site to explain why there should not have been a great many of these so-called prepottery Neolithic towns. One reason Jericho may now appear unique is that in many village sites in the Near East excavators stopped when they reached the bottom of pottery bearing deposits—pottery being equated with food production. Sometimes excavations halted because ground water was reached. This would be particularly important in erasing evidence at sites in tributary valleys of the Tigris and Euphrates. And, as W. F. Albright, the doyen of Near Eastern archaeologists, has noted, sometimes "the excavator's money and time were exhausted before he reached the bottom even of his pottery bearing stratification." [24]

The contrast between Jericho and the modest villages like Jarmo in the hill country on the margins of the Mesopotamian alluvial plain has led Robert J. Braidwood to conclude that Jericho's economy must have rested on something other than farming. He suggests that trade may be the explanation for Jericho.[25] Braidwood's reasoning is strongly influenced by his assumption that "incipient cultivation" must have taken place in the hill habitat zones of the Near East, e.g., in Kurdistan. Certainly, in Braidwood's view, farming could not have been practiced in sites such as Jericho that required irrigation at such an early period. It was in late Neolithic and Chalcolithic times in the late fourth and third millennium, argues Braidwood, that irrigation on any significant scale first became possible. Albright disagrees, for to him Jericho is evidence that the peoples of the prepottery Neolithic knew the techniques of building massive structures and that there is therefore no reason to rule out the possibility that people in the more favored piedmonts of the Near East built irrigation works. Indeed, argues Albright, it is impossible to understand the tremendous expansion of irrigation in the Chalcolithic unless we suppose that it had a long prehistory.[26]

While mortars and pestles suggest cultivation, there are no com-

[24] W. F. Albright, "Prehistory," in *At the Dawn of Civilization, A Background of Biblical History*, ed. E. A. Speiser (New Brunswick, N.J.: Rutgers University Press, The World History of the Jewish People, First Series: Ancient Times, 1964), I, 74. Comparable cultures outside Palestine started later and reached the Neolithic more than 3000 years after the Palestinian cultures. Cf. R. Vaufrey, *La Prehistoire de l'Afrique* (Paris: Librairie Masson et Cie., 1955), Vol. I, Le Maghreb, 291–368, and Anati, *Palestine*, pp. 146–48.

[25] Robert J. Braidwood, "Jericho and its Setting in Near Eastern History," *Antiquity*, XXXI (1957), 73–81.

[26] Albright, "Prehistory," p. 75.

parable implements to attest to the earliest domestication of animals. Evidence for traps designed for specific animals is found from the early Mesolithic, so that at least from this period it was practicable to capture animals without killing or severely wounding them. In Natufian sculpture young ruminants were often represented and their posture suggests to Anati that the artists observed them from very short distances. In subsequent, similar Neolithic art domestic animals were clearly portrayed, but the question of whether the earlier art showed domestic animals cannot be answered.[27]

In the area that had been occupied by Natufian and similar cultures, and which was to become the nuclear region of a Neolithic plow and seed farming complex extending from Europe in the west to China in the east, and southeast through India into Southeast Asia, goat, sheep, cattle, and pig were first domesticated. This list does not exhaust the stock of domesticated animals that appear in the area at the end of the Neolithic and Chalcolithic–early urban period. But the other animals—whether of temporary significance like the onager, or of more lasting economic significance like the ass, zebu, water buffalo, and horse—appear much later as clearly domestic in the osteological record. Domestic goats appear before 7000 B.C. in Iranian Khuzistan, Iraqi Kurdistan, Judea, and Transjordan.[28] While the osteological evidence at the earliest sites is not incontrovertible, the high proportion of yearlings in the Iranian finds suggests domestication. In Judea and Transjordan the absence of true wild goats is a strong indication that the bones found are those of domestic goats.[29]

The earliest domestic goats (and/or sheep) in Africa are found in Cyrenaica dating to ca. 4800 B.C. They apparently were introduced from Asia. In the Nile Valley and in the Fayum, however, they do not appear until much later.

Domestic sheep may be the earliest domestic animals. At Zawi Chemi Shanidar, in Kurdistan, they are dated to ca. 8800 B.C. At other sites, however, domestic sheep appear no earlier than the earliest domestic goats.

The earliest domestic pigs to have been found are at Çayönü in Anatolia, dating to early in the seventh millennium B.C. A few centuries later, ca. 6500 B.C., there is evidence of the domestication of pigs at Jarmo. Although not at first numerous, pig bones become an important component of the domestic fauna of Iraq from the middle of the fifth to the end of the fourth millennium B.C. Pigs were present in large numbers in northern Egypt in the mid-third millennium B.C., but neither cultural

---

[27] Anati, *Palestine*, p. 159.

[28] Charles A. Reed, "The Pattern of Animal Domestication in the Prehistoric Near East." In Peter J. Ucko and G. W. Dimbleby, eds., *The Domestication and Exploitation of Plants and Animals* (Chicago: Aldine Publishing Company, 1969), pp. 361–80.

[29] The ages for goats, as well as the following ones for sheep, pigs, and cattle are based on Reed, "Pattern of Animal Domestication," pp. 371–72. See also references in Reed.

nor osteological proof that they were domestic exists prior to Dynastic times.[30]

The oldest sites where remains of domestic cattle were found are at Argíssa Magula Thessaly and at Nea Níkomedeía in Macedonia. The bones are dated to ca. 6500 and ca. 6100 B.C., respectively. Subsequently domestic cattle appear in sites from Turkestan to Anatolia. Charles A. Reed remarks,

We do not know whether cattle were first domesticated in south-eastern Europe and then moved as domesticants to Asia, or whether the pattern was in the reverse direction and we have not found the early sites in Asia as yet, or whether independent centres of domestication of cattle occurred, sparked perhaps by people who travelled in trade and saw what others were doing.[31]

By the middle of the fifth millennium a full blown "neolithic" peasant complex had emerged in the Near East.[32] Since archaeological evidence for the antecedents of this complex has only recently been uncovered, it is interesting that in 1931 O. Menghin postulated that the Neolithic must have been preceded by a society that combined animal herding with grain raising. Indeed the term proto-Neolithic, which Kenyon has brought into common use, was coined by Menghin to describe the intermediate culture whose existence he reconstructed from such data as were available at the time. Menghin believed his proto-Neolithic people were sedentary, but that they may have also practiced a seasonal transhumance in search of pasture.[33]

Assuming that the proto-Neolithic settlement patterns of the Near East were those of grain raising and animal herding cultures, combining villages with seasonal camps, the reason for the emergence of the complex can be explained in different ways. Perhaps the domestication of both plants and herd animals took place within the Near East. Or it can be argued that while the evidence is strong that plants, or at least cereals, were first domesticated in the Near East, herd animals were domesticated by hunters in the Eurasian steppe, who then transmitted their techniques to the Near Eastern cereal cultivating cultures. The finds of Carleton S. Coon at the Belt and Hotu caves on the margins of the South Caspian plain are particularly interesting, for Coon claims that evidence for herding exists there prior to any evidence for harvesting of grains. Perhaps, then, herding did not develop along with or out of an agricul-

---

[30] Charles A. Reed, "Osteological Evidences for Prehistoric Domestication in Southwestern Asia," *Zeitschrift für Tierzüchtung und Züchtungsbiologie* 76, No. 1 (1961), 31–33.

[31] Reed, "Pattern of Animal Domestication," 373.

[32] Dyson, "Archaeology and the Domestication of Animals," pp. 662–63.

[33] Oswald Menghin, *Weltgeschichte der Steinzeit* (Wien, Austria: A. Schroll & Co., 1931), p. 305. See also Karl J. Narr's discussion: "Archaeologische Hinweise zur Frage des ältesten Getreideanbaus und seiner Beziehungen zur Hochkultur und Megalithik," *Paideuma* VI, No. 4 (1956), 244–50. On seasonal hunting villages see Dexter Perkins, Jr., and Patricia Daly, "A Hunters Village in Neolithic Turkey," *Scientific American*, 219, No. 5 (1968), 97–105.

tural society. At the lowest levels of the cave Coon identifies a "Seal Mesolithic" followed by a "Gazelle Mesolithic," indicating the herders were preceded by higher hunters. The geographic location of the cave is significant, for the region in which it lies opens in the east to the Turkoman steppes and in the northwest to the Black Sea region via the Transcaucasian trough. Coon therefore does not consider it an integral part of the Near East but a climatically well endowed southern extension of the great Eurasian grasslands.[34] H. Pohlhausen concludes that the domestication of herd animals was initiated by hunters who migrated along with "their" wild herds within the Aralo-Caspian grasslands and their southern bordering ranges.[35]

How then can the fragmentary and sometimes conflicting evidence of the archaeological finds best be interpreted on the basis of present knowledge? The finds of Kathleen Kenyon at Jericho, Jean Perrot at Einan, and James Melaart in Anatolia have necessitated a complete re-evaluation of long accepted theories, and it may well be that new finds will discredit interpretations based on what is known today. In a period of archaeological discoveries of revolutionary impact, no more than tentative hypotheses can be made from existing evidence. As stated earlier, archaeology has inevitable limitations, and by itself cannot provide certain evidence of the origins of phenomena. With these qualifications in mind, the following seems to be the best that can be said at present. The oldest archaeological evidence for the domestication of both plants and animals is Near Eastern. It is possible that animals were also and independently domesticated by "Aralo-Caspian" or other hunters, but if so, no acceptable evidence has yet been offered. Belt Cave and similar sites, assuming that the early finds there indicate the presence of domestic animals (Reed is doubtful about the validity of the data [36]), do not necessarily establish an independent nuclear area of domestication through a hunting culture. The general artifacts of Belt Cave point to a Natufian culture, suggesting the connection of the site to the Levant rather than to Eurasian grasslands. Menghin has provided the basis for possible classification of sites that are heavy on domestic animals and light on agriculture by characterizing

[34] Carleton S. Coon, *Cave Explorations in Iran in 1949* (Philadelphia: University Museum, University of Pennsylvania, 1951), 43–55 deal with the fauna; "Excavations in Hotu Cave, Iran, 1951: A preliminary report," *Proceedings of the American Philosophical Society*, 96, No. 3 (1952), 231–49. See also *The Seven Caves*, pp. 128–216.

[35] *Nachweisbare Spuren des Wanderhirtentums in der südkaspischen Mittelsteinzeit* (Lund: Berlingska Boktryckeriet, 1954). Pohlhausen's conceptions are elaborated in his *Das Wanderhirtentum und seine Vorstufen* (Braunschweig: A. Limbach, 1954). This view is of course in accord with Wilhelm Schmidt's theory mentioned above p. 27. However, both reindeer and horse which are often cited as primary nomadic domestications are relatively late secondary domesticates. Cf. Wolf Herre, *Das Ren als Haustier, Eine Zoologische Monographie* (Leipzig: Akademische Verlagsgesellschaft Geest & Portig, 1955) and pp. 92 and 100 below. Excellent reviews of the genesis and diffusion of Old World Nomadism are the studies of Xavier de Planhol, "Nomades et Pasteurs," in *Revue Geographique de l'Est*, No. 3 (1961), 291–310; No. 3 (1962), 295–318; and with Michel Cabouret in No. 3 (1963), 269–98.

[36] Reed, "Archaeological Evidence," p. 132 and *passim*.

them as pastoral camping sites of grain raising cultures that practiced seasonal nomadism.

Wherever domestication originated, one thing is clear: the first integration of domestic plants and animals to form a peasant complex is in the Near East. Indeed, it is questionable if such integration has ever occurred outside of the plow realm whose spread can be traced directly from its Near Eastern hearth. It is in the Near East that cattle were first used as a source of animal power; it is there that sledge, wagon, yoke, and plow are first found. The basic implements of Near Eastern agriculture are important not only for the light they throw on the integration of domesticates in the Near East to form the plow peasantry complex, but also because, as the next chapter will show, their development throws light on the relationship of Near Eastern agriculture to tropical agriculture on the one hand and to the pastoral nomadic realms of Eurasia on the other.

CHAPTER 3 *what implements show about domestication*

The relationship of tropical agriculture to the early Near Eastern peasant and plow complex has long perplexed historians of agriculture. Carl O. Sauer argues for a Southeast Asian origin of agriculture, and believes that agriculture spread to southwestern Asia and the Near East, there to become seed agriculture.[1] Archaeology, suggesting a Near Eastern origin, is held to be an unreliable guide, for the paucity of ancient remains in the tropics might be explained by the deterioration of materials. However, conditions in areas intermediate between the contesting hearths are favorable to the preservation of archaeological materials, and still none of the sites excavated (many of them down to sterile ground) reveal agricultural occupations which can even remotely compare in age with those of Near Eastern sites.

Thus the argument for tropical priority in domestication, with subsequent dispersal of domesticates or the art of domestication to Near Eastern centers, relies on a leap-frogging process to explain how agriculture appeared much later in areas that are close to the tropics than in areas that are more remote from them. Moreover, even in the Southeast Asian monsoon tropics there are numerous areas that are favorable to the preservation of archaeological materials (e.g., the Burmese dry zone and parts of Indochina), and there too intensive archaeological programs failed to yield evidence for the chronological priority of tropical domesticators.

Prehistorians generally agree that cultural connections between the tropical areas of the Old World and the Near East are much older than the period of incipient domestication. Indeed there are indications

[1] *Agricultural Origin and Dispersals* (New York: The American Geographical Society, 1952), *passim.*

that an Upper Paleolithic Eurasian hunting complex characterized by ("Gravettian") blade, graver, and microlithic industries extended into tropical Asia and Africa. In all these areas the industries are associated with Australoids or "Australiform" human types. Perhaps Australoid hunters carried their hunting and food collecting techniques from an unknown Eurasian center to Africa, southeast India, the Malayan archipelago, and beyond into the Pacific, including Australia. In all these areas Australoids have survived as more or less isolated populations, but hybridizing in Africa with yet older populations to whom they brought advanced hunting techniques.[2] That these populations, though widely dispersed, are in fact related and not merely nonrelated archaic "Australiform" types is supported by cultural similarities between contemporary Australoids and their supposed Upper Paleolithic ancestors. Thus rock paintings similar to those of modern Bushmen can be found in Smithfield and Wilton rock shelters and caves in southern Africa and similar art can be traced from western Europe by way of the Mediterranean and the Sahara through east Africa. There is a remarkable homogeneity of motifs which include hand stencils with mutilated fingers, found from Gargas and El Castillo in the Pyrenees to Africa and Australia, and painted spirit travel maps found in Australian ritual as well as in European Aurignacian and Magdalenian art. Similarly, the central Australian *churingas* or "bull-roarers" have striking counterparts in the Magdalenian and Aurignacian of Europe.[3]

Of other similarities, the one which is perhaps of greatest interest, in that it connects still current ritual practices in some areas of tropical agriculture with the era of incipient domestication in the Near East, is the treatment and preservation of skulls. Kathleen Kenyon describes the practice of Jericho:

The lower part of these skulls had been covered with clay plaster, moulded into the likeness of individual human features. . . . The eyes are inset in shells. In the case of six of the heads, the eyes are made of ordinary bivalve shells, with a vertical slit between two sections giving the appearance of the pupil. The seventh had cowry-shells. . . .[4]

The same kind of skull treatment has been found by J. Mellaart at Haçilar in Anatolia in a prepottery Neolithic culture resembling that of Jericho. And in the eastern Pacific, the excavations of Robert Suggs in

---

[2] On the historical and present day distribution of Australoids see David J. de Laubenfels, "Australoids, Negroids, and Negroes: A Suggested Explanation for their Disjunct Distributions," *Annals,* Association of American Geographers 58, No. 1 (March 1968), 42–50. On their relationship to Upper Paleolithic Hunters see Narr, "Das höhere Jägertum," 505–6.

[3] Cf. Herbert Kuhn, *On the Track of Prehistoric Man* (London: Hutchinson & Co. (Publishers) Limited, Arrow Books, 1958), pp. 46–68. Jacquetta Hawkes, *Prehistory,* I, Part 1 of History of Mankind (New York and Toronto: A Mentor Book, The New American Library, 1965), pp. 102–5, 111–12, 185–88, 293.

[4] Kenyon, *Archaeology,* p. 52.

the Marquesas have yielded comparable preserved skulls in family shrines dating between ca. 500 and 1000 A.D.[5] Actually the ritual preservation of skulls goes back further than Jericho. The oldest finds are also Near Eastern, from Erq el-Ahmar and Einan in Palestine.

Collecting skulls was a common practice in prehistoric times, and there are European Mesolithic parallels to the Natufian practice. Interpretations of the practice vary, with some arguing that it means head hunting, others that it relates to ancestral cults.[6] Whether we deal here with a phenomenon of diffusion or convergence of culture traits can only be judged on the basis of probability based on other evidence. Scholars such as F. Speiser and A. Jensen believe that there are far reaching parallels between Melanesia and, for example, ancient Greece, whether these are the result of contact, of elementary notions appropriate to an early stage of economic development, or of common derivation from an ancient cultural substratum.[7]

The megalithic monuments of the Neolithic and perhaps even of the prepottery Neolithic of the Levant have their counterparts in southeastern Asia. Megaliths were often used in the Levant for tombs, with large stones arranged to form a table-shaped structure. Frequently the stones weighed several tons and came from considerable distances. In the Palestinian megalithic assemblages there are gallery graves, dolmens, and circles. The large pit burial of Einan, surrounded by a ring of stones and covered with a slab is probably the oldest known megalithic burial in the world. It certainly is much older than similar burials elsewhere.[8] Many of the monuments in Palestine have cup marks cut into the roof slabs. Others have square or oval openings cut into the side supports.[9] In 1928 Robert Heine-Geldern argued that the megalith complexes of western and southeastern Asia and of Europe were related, and reveal that a so-called megalith culture characterized the entire area before the emergence of the urban cultures of the Neolithic–Chalcolithic. In all three areas he finds that the megaliths are similar, in both style and purpose. Everywhere the cup marks are found as well as the openings cut into the stones, presumably for the spirit to pass through, and everywhere too the megaliths seem to be graves, grave monuments, or "witness markers" commemorating some event. Heine-Geldern believes that the culture spread from the Mediterranean to both Europe and India and further east. The practice of erecting megaliths is still widespread in southwestern and southeastern Asia, and he noted particularly active

[5] J. Mellaart, "Excavations at Haçilar; Fourth Preliminary Report, 1960," *Anatolian Studies* (1961), p. 73. Robert C. Suggs, *The Hidden Worlds of Polynesia* (New York: Harcourt, Brace, & World, Inc., 1962), pp. 93–94.

[6] Cf. Anati's discussion on skull pits in Palestine and Europe in *Palestine*, pp. 172–78.

[7] Felix Speiser, "Die eleusinischen Mysterien als primitive Initiation," *Zeitschrift für Ethnologie* 60 (1928), 366, 371–72. A. E. Jensen, *Das religiöse Weltbild einer frühen Kultur*, Studien zur Kulturkunde, Vol. 9 (Stuttgart: August Schröder Verlag, 1949), pp. 66–77.

[8] J. Perrot, " 'Eynan ('Ein Mallaha)," *Israel Exploration Journal* 7, No. 2 (1957), 126.

[9] Anati, *Palestine*, pp. 278–83.

megalith cultures on Nias Island, Sumba, Flores, and in Celebes.[10] The fact that the earliest known megaliths come from the Near East, and the greater age of Palestinian megaliths compared to those of Europe or southwest and southeast Asia, as well as the similarities in construction and detail, are strong indications of actual migratory spread of Mesolithic Near Eastern peoples into the tropics as well as into Europe.[11]

But even if it were possible to establish how the megalithic structures and the methods of skull preparation spread from one area to another, the origins of domestication would be little clarified, since it cannot be known how closely these features were related to economic traits and whether all were transferred together or not. Establishing that these practices were intimately associated with plant and/or animal domesticating cultures, would not rule out the possibility that domestication was invented by a society that simply happened to have megaliths or had obtained them after the invention of food production.

While skull preparation, megaliths, and the domestic plant or animal have no obvious necessary link to one another, this is not the case with the domesticate and the agricultural tool. If the source of Near Eastern agriculture was tropical, then the characteristic tools of the earliest Near Eastern agriculture should be similar to those of tropical planting cultures, and it should be possible to show that agricultural implements found later in the Near East are merely technological advances upon the earlier tropical tools. It is evident that a mere similarity of early agricultural implements in two distant areas does not prove derivation of one from the other, for similar tools might have been invented independently in the Near East. On the other hand, if it can be shown that Near Eastern tools did not derive from tools found in tropical agriculture, then a tropical source for Near Eastern agriculture becomes very doubtful.

The characteristic tool of the Near East is the plow; of the tropics, the hoe and digging stick. Ethnologists have generally held that the plow is derived from the hoe or digging stick, an assumption which played a great, though often unconscious role in locating the invention of agriculture in the tropics. Most students today believe that it is impossible for the plow to have developed from the hoe, since the tools are operated in different ways. Hoes are employed with a forward arching movement, and the broken soil is dragged backward, the cultivator going forward from the worked to the unworked area. The plow, however, is dragged forward to make a furrow. The digging stick, like the plow, is often used to scratch the earth, but the motion is backward from the worked to the

[10] Robert Heine-Geldern, "Die Megalithen Südostasiens und ihre Bedeutung für die Klärung der Megalithenfrage in Europa und Polynesien," *Anthropos*, XXIII (1928), 276–315.

[11] William Foxwell Albright, *The Archaeology of Palestine* (Harmondsworth, Middlesex: Penguin Books, 1949), 63–64, and *From Stone Age to Christianity*, Second Edition (Garden City, New York: Doubleday & Company, Inc., 1957), pp. 136–37. Karl Narr however doubts whether Megaliths can be used as indicators of cultural connections. See "Archaeologische Hinweise zur Frage des Ältesten Getreideanbaus und seiner Beziehungen zur Hochkultur und Megalithik," *Paideuma*, VI, No. 4 (1956), 248–50.

unworked area. The same is true of the spade which Ulrich Berner, for example, believes is unquestionably derived from the digging stick.[12] Thus the presumption of the digging stick's paternity for the plow is stronger than would be that of the hoe, but there remains the contrast in the direction of movement, for one moves backward with the digging stick and forward with the plow.

Paul Leser, in a classical monograph,[13] identified a primitive plow as the oldest form. This plow is called by Fritz Kramer the J-shaped plow and by others the digging stick plow.[14] Plowhead and stilt are one piece, with the beam extending forward from it. These are found in ancient Egypt and Babylonia and are still widely distributed over the realm of plow agriculture. A consideration of the simplest J-plows, in which a rope takes the place of the beam, leads Paul Leser to the pull-spade as an antecedent of the plow.[15] Pull-spades are distributed in enclaves throughout the plow realm, particularly in Korea, the Hindu Kush region, Armenia, the Alps, and Yemen. The tools are similar to spades, but are operated by being pulled forward with ropes by hand. All that seems necessary to convert this instrument into a rudimentary plow is to replace the ropes with a beam.

But there are difficulties in accepting this instrument as the antecedent of the plow. Ulrich Berner has pointed out that wherever the tool is found, its share part is metal or metal-clad and the tool will not work without the metal sheathing. To postulate an iron tool as antecedent to wooden plows is patently absurd. Furthermore, pull-spades are operated like regular spades, with alternative strokes set next to each other and the blade pulled out of the ground between successive strokes rather than drawn through the ground continuously as in the case of the plow. What seems more likely, then, is that the pull-spade is a simplified adaptation of the plow or spade for broken and stony terrain. An impasse is reached, for none of the hand operated field or gardening tools appears a plausible antecedent for the plow.[16]

Fritz Kramer has attempted to resolve the dilemma by suggesting that the plow may not derive from any one tool, but from observation of several of them. To Kramer, China becomes the likely source for the first plows, because from a very early period there is evidence that furrowers, pull-spades, and animals were used in farming. Kramer suggests that the three were in some way combined to make an animal-pulled plow.[17] If Kramer is right, the origin of the plow cannot be tropical cultivating complexes, since neither furrowers nor pull-spades are Old World tropical tools. Yet the difficulties of deriving the plow from a single known ante-

---

[12] "Die Entstehung des Pfluges und damit zusammenhängende Probleme," *Studium Generale*, 11, No. 6 (1958), 365.

[13] *Entstehung und Verbreitung des Pfluges*, Anthropos Ethnologische Bibliothek, III, No. 3 (Münster in Westfalen: Aschendorffsche Verlagsbuchhandlung, 1931).

[14] Fritz L. Kramer, *Breaking Ground*, Sacramento Anthropological Society Paper 5 (Sacramento, California: 1966), p. 86.

[15] Leser, *Entstehung und Verbreitung des Pfluges*, p. 55.

[16] Berner, "Entstehung des Pfluges," pp. 367–68.

[17] *Breaking Ground*, p. 83.

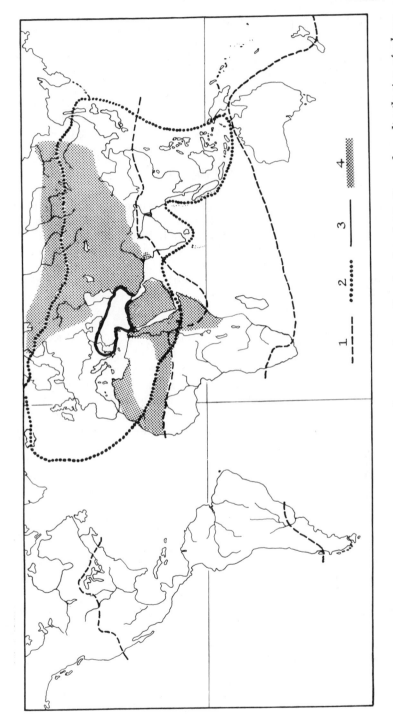

Figure 3. *Limits of Plowless Agriculture. 1, Hoe, digging stick or other hand tools; 2, Premodern distribution of plows; 3, Area of oldest plant and animal domestications; 4, Main areas of nomadic pastoralism.*

cedent tool are not cleared away by providing more than one ancestor. Neither furrower nor pull-spade is worked like the plow, for neither instrument moves in the direction that the plow does. The name "furrower" is deceptive: the tool does not make a furrow, but a groove or rill.

A further argument against tropical antecedents for the plow comes from Emil Werth's analysis of construction techniques of agricultural implements. He points to the hafting of the plow, noting that eye-hafting in which a hole is drilled in one part (whether hoe blade or plow stilt), and a handle or plow beam is inserted, is characteristic only of the "plow realm" and that "eye-hafting" goes back within that realm to Mesolithic times.[18] Kramer has pointed out that eye-hafting is not limited to areas of plow culture, but is found among the Bushmen, for example, whose grubbing sticks are weighed with drilled stones. Moreover, other methods of hafting are known in the realm of plow agriculture, for example in North Africa and India; and in Indonesia eye-hafting is entirely absent.[19] Nonetheless, the original argument holds some interest, for eye-hafting is most characteristic of the oldest area of the breaker plows (Kramer's "crook plows") which centers on the Near East. Is it possible that J-plows are more recent than breaker plows and their primitive character is the result of the simplification of the breaker plow when it was introduced into areas where physical and social conditions were not suitable for the more sophisticated instrument? In any event, among people in so-called "tropical agriculture," the method of eye-hafting is unknown.

Until recently it was assumed that the digging stick, and after it, the hoe, were the oldest agricultural implements, chiefly because of the relative simplicity of these tools. While the antiquity of the hoe has not been disproven, it has recently been questioned. Stone objects that had been identified as hoes by archaeologists or ethnographers have been shown by G. Höltker not to be hoes at all.[20] In fact Höltker's conclusion, in which Kramer concurs, is that stone was never used as a material for hoes (or plowshares). The minimum required to accept controversial stone artifacts as hoes is demonstration by ethnography that such "hoes" were hafted and used like hoes in cultivation, but such evidence has never been produced.[21] If there is no evidence of ancient stone hoes, what is the evidence for hoes made of other materials? If the oldest hoes were of wood all traces may have vanished. But there really is no evidence for the antiquity of hoes except their prevalence in tropical areas. The hoe's antiquity is a link in a chain of circular reasoning. Since tropical agriculture is assumed to be the oldest agriculture, hoes are assumed to be of great age. And since hoes are assumed to be very ancient, Old World

---

[18] *Grabstock, Hacke, Pflug: Versuch einer Entstehungsgeschichte des Landbaus* (Ludwigsburg: Verlag Eugen Ulmer, 1954), p. 146.

[19] *Breaking Ground*, p. 83.

[20] Georg Höltker, "Steinerne Ackerbaugeräte," Ein Problem der Vor- und Frühgeschichte in Völkerkundlicher Beleuchtung, *Internationales Archiv für Ethnographie* XLV, No. 4–6 (1947), 77–156. See also the comments by Carl Whiting Bishop, "Origin and Early Diffusion of the Traction Plow," *The Smithsonian Report for 1937*, The Smithsonian Institution (Washington, D.C.: Government Printing Office, 1938), pp. 531–32.

[21] *Breaking Ground*, p. 79.

tropical agriculture, where they are found, is assumed to be very ancient. Heinz Kothe has argued that hoes are recent and that there is no evidence for pre-Metal Age hoes.[22]

Many who argued for the antiquity of the "tropical hoe" felt that the priority of tropical agriculture could be established by identifying the hoe as a necessary intermediate tool between digging stick and plow. Yet it is no easier to derive the hoe from the digging stick than the plow from the hoe. The whole notion that hoe agriculture is more advanced than digging stick agriculture must be discarded, and could only have been advanced through comparison of the poorest examples of digging stick agriculture with the most advanced examples of hoe agriculture. The more "advanced" hoe is not irrevocably connected with irrigation, terracing, and manuring, although the familiar argument that it is has recently been advanced again by Sigrid Hellbusch.[23] Nor is the humble digging stick only characteristic of areas of shifting agriculture on poor soils. Digging sticks were the main implement of the intensive cultivation of Inca Peru, and hoes are used widely in Africa in shifting cultivation systems without manure or irrigation. Nor does a distinction between the tropical digging stick and the subtropical hoe constitute a valid division between hoe and digging stick agriculture, for the digging stick was found outside the tropics in New Zealand and in the American Southwest. There is indeed no reason to believe that the digging stick is peculiarly suitable to the tropics. Ulrich Berner has pointed out that the digging stick is best suited for either sticky loam soils or grassland soils with dense sod or root network.[24] If considerations of agricultural efficiency are given weight, an extratropical origin for the digging stick is clearly possible.

The consideration of agricultural implements can serve as a basis for criticizing Father Schmidt's sociological domestication thesis of a female origin for planting, i.e., domestication of plants. For societies in which women do most of the field work are much more closely connected with hoe cultivation than with digging sticks. Fritz Kramer notes: "Only in rare instances do women use the digging stick; this fact was already recognized by Hahn." [25] If the hoe is late, then "matrilineal agriculture," too, may be a late development in the history of agriculture. Unless it can be substantiated that the hypothetical female inventors of agriculture gave up digging sticks and replaced them with hoes, while elsewhere male cultivators were stuck with the digging stick, the female origin of agriculture remains problematical.

There is nothing to suggest that either the hoe or the digging stick is derived from tools first used in tropical cultivation. Digging sticks al-

---

[22] "Völkerkundliches zur Frage der neolithischen Anbauformen in Europa," *Ethnographisch-Archäologische Forschungen*, I (1953), 28–73. See conclusion on p. 73.

[23] "Vergleiche zwischen Grabstockbau und Hackbau," in *Beiträge zur Gesellungs- und Völkerwissenschaft; Professor Dr. Richard Thurnwald zu seinem achtzigsten Geburtstag gewidmet* (Berlin: Verlag Gebr. Mann, 1950), pp. 136–58.

[24] Berner, "Entstehung des Pfluges," pp. 371–72.

[25] *Breaking Ground*, p. 84.

most certainly developed from grubbing sticks of food collecting peoples, and both are more suited to grasslands than to areas of tropical or monsoonal vegetation. Indeed digging stick weights have been found in Natufian industries.[26]

The transition from grubbing stick to digging stick, in one sense, is the most remarkable revolution in the technique of food acquisition, for over a long period there was no change in the implement itself. All that was changed was the use to which it was put. Instead of being used only for digging up roots, etc., it was used to break the soil for sowing and planting. I believe that this did not happen in the tropics but in the Near Eastern sector of the Natufian cultures. An interesting indication supporting this view is that the first technical change which gave the undifferentiated grubbing/digging stick features that made it unmistakably a digging instrument resulted in what Kramer calls protospades of which numerous varieties, such as the Scottish *cas-chrom* or the Irish *loy,* are known.[27] These implements are geographically grouped around an old Near Eastern center and have attained a wider distribution than either hoes or plows. The distribution of these implements, if we follow Griffith Taylor's "zones and strata" technique, would therefore support their antiquity, for the protospade has reached further than any other agricultural implement.[28] Where the limits of the protospade coincide with those of hoes or plows, as is true in Peru of the hoe and in China of both hoe and plow, what may have happened was that the protospade was introduced with the other tools at a comparatively late period.

None of the hand (or hand and feet) manipulated agricultural implements is satisfactory as a precursor of the plow, except possibly the digging stick. If it was the digging stick, the ancestry is remote and the intermediate tools can only be surmised. Perhaps the direct antecedent was not a cultivating implement at all, but a ditching tool. Some other simple tools and techniques of agriculture were originally designed for purposes other than those with which they finally became associated. Sickles were presumably weapons before they were used for the harvest of grain. It may be that the transition from a tool whose operator steps backward (e.g., furrower) or sideways (e.g., pull-spade) to an implement which is dragged forward (the plow) came with a "protoplow" devised originally for the nonutilitarian purpose of delimiting sacred boundaries, which was done with the utmost care and awe. The sacred origin of the

---

[26] Anati, *Palestine,* p. 272.

[27] *Breaking Ground,* p. 90. On the cas-chrom and similar European tools consult Ragnar Jirlow and Ian Whitaker, "The Plough in Scotland," *Scottish Studies,* 1 (1957), 71–94. See especially pp. 71–74. See also Ian Whitaker, "Some Traditional Techniques in Modern Scottish Farming," *Scottish Studies* 3, Part 2 (1959).

[28] See Taylor's description of his concept in "Racial Geography," Chapter XIX in Griffith Taylor, ed., *Geography in the Twentieth Century* (New York: Philosophical Library, London: Methuen, 1951), pp. 433–62. According to this concept the earliest and most primitive form will occupy a zone furthest from the "cradleland." More recent forms may have obliterated the earliest form in the "cradleland" but will not as yet have suppressed them in the furthest reaches of their distribution. Taylor's concept is clearly no universal key to the diffusion of artifacts and techniques, but it is suggestive here.

plow is a prominent feature in the mythology of most ancient plow peasantries.[29] Or perhaps the "protoplow" was designed to open irrigation or drainage ditches where proper alignment was important.[30]

Recent archaeological work has established that irrigation and flood protection works are vastly older than was formerly thought. In William F. Albright's opinion, there is no basis for the view that the prepottery Neolithic knew no highly developed irrigation, but only the primitive use of springs and brooks. "While the people of Jericho did not need to build massive stone storage or deflector dams . . . they did build remarkably massive fortifications, so we may rest assured that their contemporaries in richer and more hospitable areas, did build massive irrigation works. . . ." Albright also points out that sites which are today too high for profitable irrigation have been the victims of intense erosion over millennia, while other sites have experienced a steady rise in alluvium and the earliest occupations are deep under ground water. It is therefore impossible as yet to find the earliest flood control and irrigation works.[31] It is possible, however, that a tool such as a "ditcher," drawn forward along the sightline of the operator, may have been used for irrigation works.

Irrigation works, whether or not their construction led to the development of the plow, may be indicators of the dispersal of cultivating peoples and their agricultural methods from the Near East into Eurasia and Africa. The origins of irrigation, to judge from ethnological materials, may lie in a predomestication period. Efforts to increase food supply by watering wild plants has led many primitive food collectors to spread water from a natural source in the dry season through the use of a simple hollowed stick or rills. The Hadendoa of the Eastern Desert of Egypt and the Bedouin of the Sinai Peninsula often dam or dyke wadis preparatory to sowing. Irrigation in the ancient Near East is closely connected with terracing. This association has been found in an area extending from North Africa through Palestine to Armenia, Persia, and Afghanistan in the east, and southern Arabia in the south.

Whatever the motivation for terracing, recent studies by J. E. Spencer and G. A. Hale suggest the diffusion of terracing in the Old World from a Near Eastern hearth.[32] In the course of dispersal terrace farming was adapted to different ecological environments and further modified by local cultural and historical events. Northern Indochina is identified as a secondary center of invention from which complex wet-terracing systems spread into southern China, Korea, Japan, the Philippines, South-

[29] Cf. Bishop, "Origin and Early Diffusion," pp. 532–35.
[30] Cf. Kaj Birket-Smith, *Paths of Culture*, p. 157: "The plow . . . has the same distribution as the swipe well . . ." Birket-Smith also refers briefly to a suggestion made by Gudmund Hatt that the oldest plows were used for making irrigation furrows. *Ibid.*, p. 159.
[31] Albright, "Prehistory," pp. 74–75.
[32] "The Origin, Nature, and Distribution of Agricultural Terracing," *Pacific Viewpoint*, 3, No. 1 (March 1961), 1–40. See also Paul Wheatley, "Discursive Scholici on Recent Papers on Agricultural Terracing and on Related Matters Pertaining to Northern Indochina and Neighbouring Areas," *Pacific Viewpoint*, 6, No. 2 (September 1965), 123–44.

east Asia, Indonesia, Ceylon, and Madagascar. Admittedly, the origins and diffusion of agricultural terraces have not been as intensively studied from the point of view of morphology and function as those of agricultural implements. Nonetheless it is almost certain that terracing was introduced from western Asia into the tropical areas of southern Asia. Whether the appearance of terracing there, together with techniques of water control, was associated with the first intrusion and settlement of plant and animal domesticating groups into Southeast Asia or whether the terracing belongs to a later cultural wave cannot be determined. The most that can be said is that the patterns of distribution of terracing, irrigation, megaliths, stone fortifications, cereal cultivation, and early agricultural implements reinforce the thesis of a primary hearth of Old World domestication in the Near East.[33]

The distribution of domesticated herd animals and the techniques and practices of animal husbandry similarly support the idea that the Near East was the primary center of animal domestication and that this domestication was carried out by farmers and not hunters. The reasons for believing that the big herd animals were domesticated by incipient farmers are as follows.

1. Most wild herd animals that have been domesticated lived in the realm of the ancient peasantry, whereas few whose range was primarily in the realm of nomadic hunters have been domesticated. No wild bovines, for example, in the range of nomadic hunters have been domesticated, in spite of the fact that some of these animals, such as bison, are easily domesticated. Furthermore, neither European elk nor African eland, both demonstrably easy to domesticate, have been domesticated by nomadic hunters. No deer or antelope species with the exception of the reindeer has been domesticated by hunters, and reindeer most probably do not belong to the group of the oldest domesticates.[34] None of the steppes of the New World gave rise to herding complexes although they were occupied by hunters and wild herd animals for prolonged periods of time. As far as is known, present-day nomads do not domesticate animals. In fact the use of animals in hunting, such as the cormorant, hawks, the cheetah, and the mongoose, is an invention of peasant cultures, whether the animals are used as decoys, as trackers, or as agents of the kill. Many primitive hunters do not even use dogs in the hunt. Bushmen and Pygmies are accompanied

[33] Cf. H. G. Quaritch Wales, *The Mountain of God: A Study in Early Religion and Kingship* (London: Bernard Quaritch Ltd., 1953). Siegbert Hummel, "Wer waren die Erbauer der tibetischen Burgen?" *Paideuma* VI, No. 4 (1956), 205–9.

[34] Indeed there is much in the history of some reindeer nomads that points to a descent from a sedentary rather than a hunting society. Cf. Kaj Birket-Smith's chapter on the Lapps in his *Primitive Man and his Ways*, trans. Roy Duffell (Long Acre, London: Odhams Press Ltd., 1960), pp. 103–40. On the domestication of reindeer N. T. Mirov, "Notes on the domestication of reindeer," *American Anthropologist* n.s., 47, No. 3 (1945), 393–407; Herre, *Das Ren, op. cit.* supersedes most of the earlier literature on the topic. Christoph von Fürer-Haimendorf also has reached the conclusion that the earlier view of the antiquity of nomadic reindeer and horse domestication is wrong. "Culture History and Cultural Development," in William L. Thomas, Jr., ed., *Yearbook of Anthropology* (New York: Wenner-Gren Foundation for Anthropological Research Incorporated, 1955), pp. 149–68.

by dogs, but these are not used in hunting and in fact seem to get in the way.[35]

2. The pastoral nomadic complexes of the Old World steppes always bordered on land areas of sedentary farmers who had the same domestic animals as the nomads. Conversely, in areas not adjacent to areas occupied by animal using peasantries no pastoral nomadism developed from hunting nomadism. The steppes of Australia are an example of an area occupied by hunting peoples who, in the absence of a peasantry, never became herders. The South African Hottentots are only an apparent exception. They originated in the east African steppes occupied by other herding peoples in close contact with farmers. The alleged complete psychological absorption of the pastoral nomad in his herd animals cited by those who believe that nomadic hunters were the first domesticators has been shown to be irrelevant. Intense psychological attachment to animals can develop in a short time. American Indian hunters became horse riding nomads shortly after the Spaniards introduced the horse into North America. Recently Homer Aschmann has described another example: in Latin America, a formerly sedentary cultivating society is now completely absorbed in pastoral husbandry.[36] In this connection it is also interesting that where domestic animals have been transmitted to hunters in historical times, these hunters have not become pastoral nomads, but have used the domestic animal in order to facilitate their hunting, e.g., the bison hunting Indians became bison hunters on horseback. That the Navaho became sheepherders under American direction does not vitiate this thesis, for game had been eradicated, the old hunting range had been severely restricted, and an alternative had been explicitly provided. All this suggests a sedentary (farming) origin not only of herd animal domestication, but even of pastoral nomadism.

3. The problem of feeding captured animals, the solution of which was a prerequisite for domestication, is believed by some students to be soluble only by an agricultural society producing food surpluses that might be used to supplement pasturing.

The techniques of animal handling used by nomadic peoples also suggest a peasant origin. As has been shown by detailed studies, the harnessing methods used by nomadic societies are modifications of those farmers devised for handling herd animals in the field. Whatever doubts exist as to who the original domesticators were, whether farmers or hunters, there is no question that cattle were originally domesticated by farmers. It is significant therefore that the oldest types of harness applied to other animals suggest that cattle were the first animals used for traction. The oldest known cattle harness is the double neck yoke which makes it possible to utilize the great muscular power in the cervical and thoracic region of cattle. Such harnessing has been found in Mesopotamia asso-

[35] There has been no adequate ethnological study of the antiquity of the dog in Pygmy, Bush- and Negrito hunting societies. It is certain that the Andaman Islanders first obtained dogs from the English. W. Nippold, *Rassen- und Kulturgeschichte der Negritovölker Südost-Asiens* (Leipzig: Jordan & Gramberg, 1936), pp. 178–79.

[36] "Indian Pastoralists of the Guajira Peninsula," *Annals*, Association of American Geographers, 50, No. 4 (1960), 408–18.

ciated first with wagons and later with plows. Before the development of the neck yoke the animals were merely tied to the plow or wagon either by ropes tied directly to the horns or by a beam lashed to the horns. After Mesopotamia, the harness appeared in other areas of high civilization in the plow realm, in Egypt as well as in northwest India. The oldest known representations, whether among farmers or nomads, show that equids were harnessed by both neck yokes and head yokes, both of which are physiologically inappropriate to the animals and could only have been transferred to them from bovine harnesses.[37] It would thus seem that the large nomadic pastoral realm of Eurasia owes its harnessing methods to the old peasantries. Even harnesses, pack-saddles, and riding saddles employed with reindeer appear to some students to have been copied from horse or cattle gear. The riding saddles of the Tungus and Sotot follow closely the Altai, Turk, and Mongol type. The pack saddles of such reindeer nomads as the Chuckchi, Koryak, and Samoyed are of the same pattern. Similarly Lapp pack-saddles and sledge harnesses are derived from harnessing methods employed by farmers.[38]

An old argument for the independent origin of nomadic domestication has been the existence of practices considered peculiar to pastoral nomads, such as presenting a cow or camel with a straw-stuffed calf or camel skin to stimulate milk flow, blowing into anal or vaginal passages, and milking from behind.[39] But these are now known to have been common in the realm of ancient west Asian peasants and presumably derive from that realm. Indeed ethnologists have found few, if any, practices of animal husbandry among nomadic pastoralists that are not of great antiquity among sedentary farmers. This applies also to other traits identified by Hatt as the earliest cultural elements of nomadism, such as marking of ears, lassoes, the use of salt to attract herd animals, decoy hunting etc.[40] While arguments from techniques are not conclusive, they point to the prior peasant domestication of the animals of nomadic pastoralists.

[37] See Erich Isaac, "On the Domestication of Cattle," *Science,* 137, No. 3525 (July 1962), 196, for references. See also Franz Hancar's comprehensive study: *Das Pferd in prähistorischer und früher historischer Zeit,* Wiener Beiträge zur Kulturgeschichte und Linguistik, Band XI (Wien–München: Verlag Herold, 1956), 414–21, 432, 436 and passim. On the diffusion of wheeled vehicles from the Near East see V. G. Childe, "The First Waggons and Carts—From the Tigris to the Severn," *Proceedings of the Prehistoric Society,* New Series, XVII, part 2 (1951), 177–94, and "The Diffusion of Wheeled Vehicles," *Ethnographisch-Archäeologische Forschungen,* II (1954), 1–17. See also Stephen Foltiny, "The Oldest Representations of Wheeled Vehicles in Central and Southeastern Europe," *American Journal of Archaeology,* 63, No. 1 (January 1959), 53–58 and A. G. Haudricourt, "Contribution à la géographie et à l'ethnologie de la voiture," *La Revue de géographie humaine et d'ethnologie,* I (1948). R. J. Forbes, *Studies in Ancient Technology,* Vol. II, 84–87 discusses the inefficient use of animal power due to the transfer of cattle harnesses to other animals.

[38] Zeuner, *Domesticated Animals,* p. 126.

[39] E. C. Amoroso and P. A. Jewell, "The Exploitation of the Milk-ejection Reflex by Primitive Peoples," *Occasional Paper No. 18 of the Royal Anthropological Institute* (1963), pp. 126–37.

[40] Gudmund Hatt, "Notes on Reindeer Nomadism," *Memoirs of the American Anthropological Association* (Lancaster, Pa.: Published for the Am. Anthrop. Assoc., 1919) VI, No. 2, 75–133.

CHAPTER 4 *the domestication*
of plants

An obvious approach to studying the domestication of plants is to discover the natural distribution of the wild ancestors of plants known to be early domesticates. The difficulty is that the phylogeny of many domesticates is in doubt. Even where the evolution of a subgenus to which a domesticate belongs is adequately known, the possibilities are merely somewhat narrowed to a range of potentially domesticable, often widely distributed plants. The student is finally compelled to look for other evidence to reveal where the original domestication occurred. The difficulties with other types of evidence have been demonstrated in the preceding chapters.

It cannot be assumed that among the potentially domestic plant races which have been mapped out, the most primitive is necessarily ancestral to the earliest domesticate. For example, *Gossypium herbaceum africanum*, which is the most primitive wild form of cotton, played no role whatever in the development of domestic cottons. It is even debatable in some instances whether the first "prodomestic" plants, those immediately ancestral to a domestic crop, grew inside the distribution of the wild ancestral plant. That they grew outside that distribution is not as implausible as it might first appear. Domestication is the result of more than the utilization of a wild plant in its natural distribution, for such utilization goes back to our prehuman ancestors. It could be argued that taking a wild plant (or animal) outside of its natural distribution was one factor in initiating a series of changes that culminated in the domesticate. Available evidence suggests wheat was domesticated in this way.

Still another problem is that what is identified today as a wild "ancestral" plant may not be a descendant of the progenitor of some domestic plant, but rather an offshoot of a domestic plant or a weed race that escaped cultivation or ran wild on abandoned but formerly cultivated land. Many of the Eusorghums of Africa may have originated in this way. The problem is analogous to one encountered in delimiting the range of prodomestic animals. There are cases where animals considered wild were in fact feral. All this does not mean that distribution of potentially ancestral plants and animals can be ignored in the study of domestication, but only that the evidence must be treated with caution. In all probability, even if the original domestication did not take place somewhere inside the range, it occurred in a peripheral zone.

The most influential modern theory concerning the geographical origin and spread of domestic plants has been that of the Russian plant geographer N. I. Vavilov. He argued that the original centers of plant domestication are those areas where the greatest number of cultivated forms of the domesticate are found.[1] This theory opened up the possibility of deducing the original centers of domestication from present-day centers of variability. The attraction of the thesis stemmed not only from the light it cast on the origins of particular domestic plants, but also from the perspectives it opened on cultural history, historical geography, and agricultural botany. Vavilov's theory offered a simple key to the most ancient secrets. No longer did the scholar have to resign himself to waiting a hundred years or so for reliable archaeological evidence. He could go out now, or send his assistants out, to collect wild and cultivated varieties of domestic plants. Yet much as the theory has contributed to geo-botanical discovery, it is an inadequate conceptual tool for the study of the origins of domestication.

Vavilov's theory locates the hearth or hearths of the major domesticates. Emmer wheat and barley are probably the oldest domestic grains, and Vavilov found that the primary center of origin of diploid wheats and barley was Abyssinia. "Barely half a million hectares are under wheat in Abyssinia . . . *Yet, according to the number of its botanical varieties of wheat Abyssinia occupies first place. . . . This is also the center of origin of cultivated barley. Nowhere else does there exist in nature such a diversity of forms and genes of barley. . . ."* [2] From the Abyssinian center, according to Vavilov, emmer spread to Yemen and via Egypt through the Levant to Mesopotamia in the east, Turkey in the west, and then on to Europe. But as Helbaek notes, Vavilov actually reversed the route indicated by the archaeological evidence. Abyssinian emmer today is of the same variety as the emmer of ancient Egypt; it seems to have been introduced into Abyssinia at a time when it was still cultivated in Egypt; and it probably spread from Abyssinia east into the Indian Ocean littoral. Since wild emmer, *T. dicoccoides*, is not present in Abyssinia, Helbaek concludes, "Vavilov

---

[1] Nikolai I. Vavilov, "Studies on the Origin of Cultivated Plants," *Bulletin of Applied Botany and Plant Breeding*, 16, No. 2 (1925), 1–248. For Vavilov's "World Centers of Origin" of cultivated plants see *The Origin, Variation, Immunity and Breeding of Cultivated Plants*, trans. K. Starr Chester, Chronica Botanica, 13, No. 1/6 (Waltham, Mass.: The Chronica Botanica Co., 1951), 20–43. Quoted henceforth as *Origin*.

[2] Vavilov, *Origin*, p. 38.

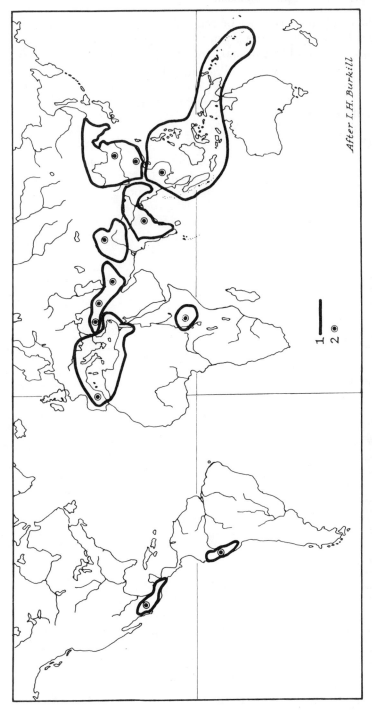

Figure 4. *Centers of Origin of Cultivated Plants, according to N.I. Vavilov.*
1, *Regions of domestication;* 2, *Main centers of variability.*

After I.H.Burkill

simply rejected it as the progenitor of cultivated emmer, *T. dicoccum.*"[3]

In the case of barley, Vavilov's "multiplicity center" or "gene center" theory is also not borne out. The discoveries of Jarmo, on the contrary, support the view that agriculture began in the Zagros–Taurus–Palestine rain-watered area.[4] Abyssinia's large diversity of wheats and barleys under relatively uniform ecological conditions in a limited terrain above 4500 feet (which causes Vavilov surprise) actually supports the thesis that a number of different domestic wheats and barley were introduced into a climatic realm somewhat similar to that of the Near Eastern centers of original domestication. Where ecological conditions differed in Abyssinia, further diversification of barley and wheat was the result. The weakness of Vavilov's conception is that he equates centers of variability with hearths of domestication, whereas in the case of domesticated plants, such centers more often occur in the frontier zones or an isolated peripheral deme of their range.[5] For example, the Swabian region of Germany was a center of spelt cultivation by 1000 B.C. and a center of diversity. Under the influence of Vavilov, botanists have claimed spelt originated in Swabia.[6] A center of diversity created by the elimination of formerly dominant plants (in this case emmer and einkorn) in frontier environment for these plants, and the survival of spelt variants which elsewhere would have been suppressed, has been misinterpreted as a center of origin.[7]

Another weakness of Vavilov's theory is that modern distributions of wild plants are often relict distributions of a former range. Also, areas with a multiplicity of wild forms are frequently remote from the range of the genetically possible ancestral plants. The centers of multiplicity of wheats, for example, are in the western Himalayan ranges, a considerable distance from the range of ancestral wild wheat.[8] Similarly, Vavilov locates one hearth of flax domestication in a central Asiatic center including northwest India, Afghanistan, Tadjikistan, Uzbekistan, and western Tian-Shan. Yet Helbaek has shown that these areas lack a possible progenitor for flax.[9] Vavilov identified the multiplicity center of bread wheats in the same area,[10] yet it is now clear that the bread wheats originated from hybridization of tetraploid wheats with the fourteen chromosome grass *Aegilops squarrosa.*[11] The only place where such hy-

[3] Helbaek, *Paleoethnobotany*, p. 102.
[4] *Ibid.*, p. 110.
[5] A pertinent essay treating *inter alia* geographic speciation is M. J. D. White's "Models of Speciation," *Science*, 159, No. 3819 (8 March, 1968), 1065–69.
[6] Karl and Franz Bertsch, *Geschichte unserer Kulturpflanzen* (Stuttgart: Wissenschaftliche Verlagsgesellschaft, 1947, pp. 39–42.
[7] Alfred C. Andrews, "The Genetic Origin of Spelt and Related Wheats," *Der Züchter*, 34, No. 1 (1964), 17–22.
[8] Wolfgang La Baume, *Frühgeschichte der europäischen Kulturpflanzen*, Giessener Abhandlungen zur Agrar- und Wirtschaftsforschung des europäischen Ostens, 16 (Giessen: Kommissionsverlag Wilhelm Schmitz, 1961), 26. La Baume cites further valid objections to Vavilov's theory on pp. 25–29 of this study.
[9] Helbaek, *Paleoethnobotany*, p. 116.
[10] Vavilov, *Origin*, p. 31.
[11] Cf. Herbert G. Baker, diagram on wheat in his *Plants and Civilization* (Belmont, California: Wadsworth Publishing Company, Inc., 1965), p. 68.

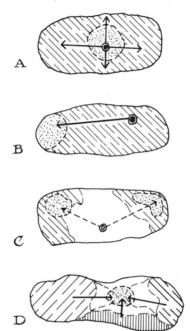

Figure 5. *Possible Relationships of Variability Centers to a Hearth of Domestication.* **A,** *Hearth of domestication (ringed dot) located inside center of variability (dotted area).* **B,** *Variability center (dotted area) at limit of distribution of cultivated plant.* **C,** *Extinct hearth (concentric circles) of domestication; crop given up in oldest areas of cultivation (white area); present cultivated area (slanted lines) and centers of variability at margins of expansion of the crop.* **D,** *Center of Variability (dotted area) as relic area: former distribution of different varieties overlapped in center of variability from which they have now receded. Evacuated area is white.*

bridization, between the useless *A. squarrosa*, which occurs as weed in wheat fields, and an emmer wheat, can occur is in the area from the Balkans to the Caucasus.

In the 1930s Elizabeth Schiemann concluded that Vavilov's theory that domestic plants may have a number of different centers of domestication is not supported by the evidence.[12] Wherever Vavilov found a multiplicity of wild varieties he concluded there was a separate hearth of domestication. But according to Schiemann, centers of multiplicity are rather areas of hybridization or areas at the limits of the species. A number of other scholars have pointed out that culture has determined the location of "multiplicity centers." Selection, increased rates of mutation, and increased potential for hybridization are brought about by man, even if sometimes inadvertently, by increasing plant population densities under cultivation. Thus while it is possible that multiplicity centers and hearths of domestication in specific cases coincide, it is neither necessary nor probable that they frequently do so. Where studies of wild plants have been made, their multiplicity centers were definitely not the center of their evolution.[13]

---

[12] For an emphatic summary of her view see "Gedanken zur Genzentrentheorie Vavilov's," *Die Naturwissenschaften* 27, No. 22 (2 June, 1939), 377–83 and 27, No. 23/24 (9 June, 1939), 394–401.

[13] One of the problems with Vavilov is that he never clarifies the difference between geographic speciation and the origin of domesticated plants. See E. Schiemann, "Genzentrentheorie," p. 399.

In the case of domestic plants, the centers of multiplicity identified by Vavilov are always areas where primitive agriculture has survived. He finds hearths of domestication in Peru and in Mexico in the areas that were the main pre-Columbian centers of agriculture and where native farming lasted longest. The centers of domestication for wheat and barley he puts in the western Himalayas and in Abyssinia where primitive farms have also survived. As Rothmaler points out, multiplicity centers of domestic plants occur where primitive farming maintained itself in what may be called cultural refuge or areas of relict distribution. These are often mountainous regions of considerable ecological diversity, and the farming cultures found there do not generally select for uniformity.[14]

Although Vavilov believed there were multiple hearths of domestication and that the same plant might be domesticated independently in different "gene centers," his theory has tended to lead scholars to concentrate upon a tropical center for plant domestication. The primitive nature of such surviving tropical agriculture had long encouraged an assumption that the tropics must be the primary area of cultivation. When this consideration was combined with emphasis on the variety of vegetative forms and their extensive use, the Asian tropics—specifically the monsoon tropics of Southeast Asia—became an obvious region to select as the primary hearth of plant domestication. The difference in method of propagation of plants in the tropics has buttressed the theory. In the tropics multiplication of plants by seeds is less important than propagation by cuttings from root or rhizome tubers or stems and vines which are buried in the soil. Carl O. Sauer, who regards the "old planters" of the tropics as the first domesticators, stresses that propagation of plants in tropical planting systems proceeds on the principle of asexual reproduction and "seeds being of no interest, many such cultivated plants have lost the capacity to bear viable seeds, some as sterile polyploids, some in other ways."[15] In other words, these plants are cultigens, or man-made plants, and such far-reaching changes as loss of viable seed production could only have been the result of long tampering with the plant.

But quite apart from the fact that numerous nondomestic plants propagate asexually, the relative importance of root and stem tubers in tropical agriculture cannot be taken as symptomatic of the priority of plant domestication in the tropics. To emphasize the difference between tropical cultigens and the domesticates of nontropical areas is to ignore the importance of roots and tubers in the ancient food budgets of the Near East. Roots and tuberous plants were gathered by the preagricultural peoples of western Asia and remained important after grain agriculture had become firmly established. What presumably happened was that roots and tubers were used by the incipient agriculturists of the

[14] W. Rothmaler, "Die Heimat unserer Kulturpflanzen," *Urania*, 12, No. 5 (1949), 15–20; Cf. Ames, *Economic Annuals*, pp. 133–35.
[15] Sauer, *Agricultural Origins*, p. 25.

Near East for various purposes and that after the domestication of grains, specialization occurred, with grains becoming the staple and roots and tubers falling to secondary use, e.g. as condiments, flavoring for beers and wines, desserts, etc. Perhaps there was a carry-over phase marked by the continuing use of wild plants after the domestication of the staples (some of these wild plants were later domesticated, others continue to be gathered today). The Swiss lake dwellers used both the starchy fruit of the water chestnut (*Trapa natans*) and its fleshy roots long after they specialized in grain raising. The plant has continued to be important among other grain raising peoples. It is not used any more in Europe or the Near East, but is common in Kashmir, and a related variety *T. bicornis* is cultivated in China and Southeast Asia. Similarly the papyrus tuber was extremely important in Egypt. Not only was the plant used for paper making, but the roots were eaten raw, or roasted or cooked in the form of pounded meal prepared as porridge. Papyrus root porridge remained until recently a favorite infant food in Egypt.

While the plants just mentioned were, so far as is known, collected rather than cultivated, plants with thickened roots, hypocotyls, stems, or shoots were clearly domesticated by grain raisers. The numerous varieties of domestic beets, ranging from stock feeding beets to sugar beets, derive from *Beta vulgaris* var. *maritima,* which is a Mediterranean plant. Carrots, too, are very ancient cultigens which have changed so radically in cultivation that it has thus far not been possible to recreate the carrot experimentally from the assumed ancestral European *Daucus carota* or the Mediterranean *D. maxima.* Parsnip, parsley, caraway, salsify, black salsify, turnips, horse radish, celery, celeriac, and gentian are some plants whose roots no less than their leaves were, and in many cases still are used. The antiquity of some of the midlatitude tubers is such that no wild occurrence is known, as in the cases of various tuberous species of *Brassica.* Others such as horseradish, *Armoracia lapathifolia,* have to be propagated by "cuttings" since the domesticated variety does not produce viable seed.

Perhaps the harvesting of grain directed attention to buried or semi-buried tubers. The diminished ability to produce viable seed, or the loss of that ability altogether, if it was the product of tampering with the plant, can be explained by the removal of flowering and leafy parts, which is what is done in the harvesting of grains. The technique might have been applied to tubers in the course of the harvesting of domestic grains, since many of the tuberous plants are intruders in grain fields, e.g. salsify, horseradish, carrots etc. Moreover, only if the flowering parts were removed would the digging up of tuberous roots have been rewarding, for many of the tuberous parts become truly globular and stocked with nutrients only after leaf pruning and removal of the inflorescences which deplete the food stored in the roots.

Plants were harvested in the tropics long before agriculture, but the transition to cultivation occurred, I believe, only under the impact of ideas and techniques which arrived from western Asia. To hold that the flow was reversed is to believe that plant domestication alone moved against the general flow of the cultural current, for the flow of Upper Paleolithic-Mesolithic and Neolithic culture in Eurasia was overwhelmingly toward rather than away from the tropics. Furthermore, had the dispersal of agriculture proceeded outward from a tropical hearth, "tropical cultures" ought to appear in the archaeological deposits of the Near East previous to the domestication of grains. There is no evidence for such a cultural intrusion. What seems far more likely is that the grass domesticating cultures introduced their crops to the tropics wherever possible. In the face of ecological barriers, the grains were given up.

Even if the thesis that the oldest center of plant domestication was tropical is rejected, the possibility remains that tropical domestication occurred independently of an older center. G. P. Murdock argues for an independent invention of agriculture in west Africa, identifying the plants of the "Sudanic complex" which he believes were domesticated by the "Nuclear Mande" peoples around the headwaters of the Niger.[16] The domesticates Murdock attributes to the Mande peoples include some that are clearly of Asian origin, e.g., cotton, sesame. Other plants of his list may have been derived from an African wild plant, but their domestication occurred after the intrusion of Near Eastern domesticating cultures. Surveying the history of cultivated sorghum, H. Doggett argues against Murdock:

To me, it seems more probable that the Cushites brought agriculture to Africa. . . . These Caucasoid people presumably passed through or near Iraq-Kurdistan on their journey to Abyssinia. . . . We may suggest that the Cushites grew emmer wheat, and being accustomed to cereals, developed some of the local plants including *Eleusine* millet, teff, flax and sorghum, for use in the area less well suited to wheat. . . . To a people practicing primitive agriculture the wild Eusorghums would be obvious grasses to adopt and improve. Indeed they might well have been weeds in their field.[17]

In other words Doggett considers African sorghums and millets secondary or substitute domesticates. A few of Murdock's plants are not even domestic. The Shea-Butter tree, *Butyrospermum parkii,* is not a domesticate, but a wild tree whose seeds are harvested from the Nile almost to the Atlantic. Two others, the fluted pumpkin *Telfairia occidentalis* and Fonio millet *Digitaria exilis,* are "either undifferentiated from wild plants still occurring in the area or else show a close and obvious relationship. They could have been added to an agriculture which came from the east."[18] The twenty-five plants on Murdock's list can be ac-

[16] George Peter Murdock, *Africa: Its Peoples and Their Culture History* (New York, Toronto, London: McGraw-Hill Book Company, Inc., 1959), pp. 64–67.

[17] H. Doggett, "The Development of the Cultivated Sorghums," Chapter III in *Essays on Crop Plant Evolution,* ed. Sir Joseph Hutchinson (Cambridge, England: Cambridge University Press, 1965), p. 59.

[18] H. G. Baker, "Comments on the Thesis that There Was a Major Centre of Plant Domestication Near the Headwaters of the River Niger," *Journal of African History,* III, No. 2 (1962), 232.

counted for by the assumption that they are either not domestic at all, substitute domesticates for plants originally introduced from another area, or outright introductions.[19]

But if original or independent centers of domestication in the tropics are thus far ruled out, we are left with the original hypothesis of a Near Eastern hearth of domestication. By the end of the eighth millennium B.C., a widespread village pattern had emerged in the Near East, its economy centering on the cultivation of cereals.[20] Assuming that this is the hearth of plant domestication, then, we will discuss the primary domesticates of this hearth, the process whereby domestication spread out from this area, and the subsequent domestication of other plants both in this area and outside of it.

BARLEY AND WHEAT. The primary domesticates of the Near Eastern hearth were two-rowed barley, emmer, and einkorn. The barley genus, *Hordeum,* is divided into two main groups: two-rowed and many-rowed barley types. The first domestic barleys of the Near East were two-rowed, *H. distichum,* and resemble the wild spontaneous occurring barleys of the region, *H. spontaneum.* Massive stands of spontaneous two-rowed barley are found today in the foothill area of the western Zagros, southern Taurus, Anti-Taurus and Anti-Lebanon ranges, the Hauran, Jebel Druze, and the Jordan drainage area, their distribution coinciding with that of early farming villages. In both the wild and domestic forms, only the median florets are fertile and develop seed. Hence only two rows of kernels develop on each spike. The major difference between wild and domestic forms is that the wild form has a brittle rachis which breaks on maturity into individual units, each containing a seed and the two lateral spikelets, while the rachis of the domesticated form is tough and does not break even on threshing. In both *H. spontaneum* and *H. distichum,* on ripening the palets and lemmas generally fuse with the kernel, producing a "hulled" grain,[21] and no amount of threshing can remove the "hull" of "hulled barley." There are, however, varieties of domestic two-rowed barley which are naked, *H. distichum* var. *nuduum;* the kernels are free and the palet and lemma scales come loose from the kernel.

The earliest indication of domestic two-rowed barley comes from Jarmo. In the sun-dried adobe that was used for buildings the imprint

[19] J. Desmond Clark, "The Spread of Food Production in Sub-Saharan Africa," *Journal of African History,* III, No. 2 (1962), 211–28. Ivan R. Dale, "The Indian Origins of Some African Cultivated Plants and African Cattle," *The Uganda Journal,* 19, No. 1 (March 1955), 68–72. Christopher Wrigley, "Speculation on the Economic Prehistory of Africa," *Journal of African History,* I, No. 2 (1960), 189–203.

[20] Cf. Kent V. Flannery, "The Ecology of Early Food Production in Mesopotamia," *Science,* 147, No. 3663 (12 March 1965), 1247–56. T. Cuyler Young, Jr., and Philip E. L. Smith, "Research in the Prehistory of Central Western Iran," *Science,* 153, No. 3734 (22 July 1966), 386–91. Frank Hole, "Investigating the Origins of Mesopotamian Civilization," *Science,* 153, No. 3736 (5 August 1966), 605–11. On Palestine and neighboring areas see Anati's outstanding *Palestine,* passim.

[21] Werner Rau, *Morphologie der Nutzpflanzen* (Heidelberg: Quelle & Meyer, 1950), p. 226.

of two-rowed barley similar to the wild species has been found. The reason for thinking that it may have been domestic is that carbonized fragments have been found that show the rachis was not brittle. As Helbaek has noted, "After carbonization at the site and after excavation millenniums later, then transport and handling in the laboratory, a brittle axis would not show unbroken joints. In contrast, a dry present day spike of wild barley, even if not fully mature, breaks into pieces when touched." [22] Today lineal descendants of the wild ancestors of Jarmo's carbonized barley still grow "in the fresh dirt of the dump heaps of the excavation." [23] Domestic two-rowed barley is not found in Europe until classical times. There is evidence for it in third century B.C. Greece and in Italy three hundred years later. In northern Europe it only appears after 1000 A.D.[24]

The origins of domestic many-rowed barleys, especially of six-rowed barley, *H. hexastichum,* have never been determined. Scholars have hesitated between the opinions of F. Koernicke, who considered *H. spontaneum* the prototype of all cultural forms; de Candolle, who entertained the possibility of a wild six-rowed barley, now extinct, as ancestral to the six-rowed groups; and W. Rimpau, who went to the extreme of deriving all barley types from six-rowed barley.[25] This was at least a reasonable conjecture until recently, because all the prehistoric barley found in Swiss lake dwellings and in Egyptian tombs was six-rowed. The view can no longer be justified now that domestic two-rowed barley has turned up about 4500 years before "the first hoe turned the soil of Switzerland" and about 4000 years before the first pyramid was built.[26] Genetic experimentation has shown that six-rowed barley may derive from two-rowed barley. Alleged wild six-rowed barleys that have been found are neither wild nor old. For example, a six-rowed barley, *H. agriocrithon,* discovered recently in east central Asia and identified as "wild" because it has a brittle rachis, is actually a formerly cultivated form which reverted to this primitive characteristic. The area where it was found is an ecological border region where many barley types were developed by cultivation, such as four-row hull-less Jerusalem barley, *H. coeleste,* and four-row hooded, or Nepal barley, *H. trifurcatum,* of which both naked and hulled forms occur. Helbaek believes that six-rowed barley first appeared as a mutant in southern alluvial Mesopotamia and Cilicia. He explains the many varieties of both-rowed and irregularly deficient two-rowed types in the archaeology of the Fayum of Egypt by postulating that the Fayum barley "is an example of the species in a

[22] Helbaek, *Paleoethnobotany,* p. 108.
[23] *Ibid.,* p. 109.
[24] *Ibid.,* p. 108.
[25] Cf. Wilfred W. Robbins, *The Botany of Crop Plants* (Philadelphia, Pa.: P. Blakiston's Son & Co., 1917), p. 148. W. Rimpau, Die genetische Entwicklung der verschiedenen Formen unserer Saatgerste. *Landwirtschaftliches Jahrbuch,* 21 (1892), 699–702.
[26] Hans Halbaek, "Domestication of Food Plants in the Old World," *Science,* 130, No. 3372 (14 August 1959), 370.

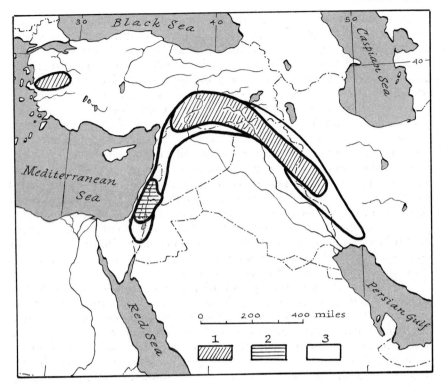

Figure 6. *Natural Distribution of Ancestral Wheat and Barley. 1, Einkorn; 2, Emmer; 3, Barley.*

state of vigorous mutation in consequence of its fairly recent introduction into the Nile Valley." [27]

Emmer, *Triticum dicoccum*, and einkorn, *T. monococcum*, the other two primary domesticates, are both types of wheat. In the archaeological record identification rests on morphological characteristics of imprints or carbonized remains. Both emmer and einkorn are "hulled," "glume" or "spelt" wheats, in which the grain remains attached to the lemma and palet and is not released when threshed. The rachis of the ripe wheat is fragile. They are distinguished from "naked wheats" where the ripe rachis is tenacious, but the grain comes free from the lemma and palet and is easily threshed. The remains of "hulled" wheat are easier to identify archaeologically, although sometimes naked wheat can be distinguished on the basis of shape.

Einkorn is particularly close to its wild form, *T. boeoticum* Boiss. emend. Schiem., the most conspicuous differences being that the rachis is not as fragile and the grain is somewhat larger. Einkorn hybridizes so readily with the wild form that in the opinion of some, wild and cul-

[27] Helbaek, "Paleoethnobotany," p. 111.

tivated einkorn should be considered one species. Harlan and Zohary report that wild einkorn is most at home on basaltic cobble in southeastern Turkey. In Turkish Kurdistan near Diyarbekir extensive rocky slopes covered with almost pure stands of wild einkorn were found in association with *T. speltoides* (*Aegilops speltoides*). These wild grasses are the parents of wild emmer which continues as a minor element in the vegetation. The extensive stands have been described by Harlan and Zohary: "Massive stands range from elevations of 2000 meters down to the edge of the plains in Urfa and Gaziantep (about 600 meters). Over many thousands of hectares it would be possible to harvest wild wheat today from natural stands almost as dense as a cultivated wheat field. If the present abundance is any indication of the situation 10,000 years ago, food gatherers and collectors would surely have been attracted to the stands in southeastern Turkey." [28] Formerly the range of wild einkorn was probably larger, and massive stands still occur outside Turkey in northern Iraq and Iran. By the third millennium einkorn is found in some quantity in Troy and over much of the Danubian area, central and western Europe, usually together with emmer.

Emmer arose from the hybridization of wild einkorn and a diploid grass *T. speltoides* (*Aegilops speltoides*).[29] Like einkorn, emmer is so similar to its wild form that it might have been grouped under the same taxonomic name, although traditionally it is called *T. dicoccum*. The main difference is that in domestic emmer the rachis is somewhat stronger than in the wild forms. On botanical grounds wild emmer may be the best indicator of the region of earliest cereal domestication, for the plant is not weedy and is demanding in its requirements. Hence it does not spread easily from established habitats. Wild emmer occurs in two ecogeographic types fairly well separated from each other. One variety ranges from southeastern Turkey through northern Iraq to western Iran with an outlier in the Caucasus. This northern "race" is never abundant, but usually occurs in isolation or in thin scattered stands associated with wild einkorn and wild barley. Wild emmer is never a dominant, and Harlan and Zohary believe that since it is not weedy the range of this cereal may have been constricted from the time the land was first cultivated.[30] The other variety, which is centered in the northern Jordan valley, is large, robust and has big seeds. It occurs in massive stands on basaltic and limestone slopes from Galilee and Mount Hermon to the mountains of Gilead and Jebel Druze. But, say Harlan and Zohary, "It was not until the nation of Israel was established and grazing was con-

---

[28] Jack R. Harlan and Daniel Zohary, "Distribution of Wild Wheats and Barley," *Science*, 153, No. 3740 (2 September 1966), 1078.

[29] The problem of classification of species within the genus *Triticum* was only solved cytologically by the Japanese scientist Sakamura in 1918. Sakamura showed that there were three groups of wheat differing in number of chromosomes: the 14 chromosome wheats (with two sets of 7) hence called diploid; the 28 chromosomes (4 x 7) tetraploid wheats, and the 42 chromosome (6 x 7) hexaploid wheats. Cf. Ralph Riley, "Cytogenics and the Evolution of Wheat," Chapter V in Hutchinson, *Essays on Crop Plant Evolution*, pp. 103–22.

[30] "Distribution of Wild Wheats," p. 1079.

trolled that the abundance of these stands was recognized. Where grazing is controlled, nonarable sites support stands as dense as cultivated wheat fields."[31]

The Palestinian wild emmer readily crosses and produces fertile hybrids with most of the present tetraploid cultivated wheats as well as with cultivated emmer. These hybrids include durum, *T. durum*, poulard or rivet wheat, *T. turgidum*, and "Polish" wheat, *T. polonicum*.[32] The Taurus-Zagros race of emmer is cytologically different, crosses poorly, and shows decided sterility with the same cultivated tetraploid wheats. Thus present evidence indicates that most of the modern tetraploid wheats stem from the Jordan valley emmer, while einkorn was probably first domesticated in southeastern Turkey. Barley could have been domesticated anywhere within the Taurus-Zagros arc or the Turkey–Palestine axis. Since the plants have somewhat different ecological requirements, barley being the most xerophytic and einkorn more tolerant of cold, it seems likely the plants were domesticated in different areas of the Near East.

Archaeological evidence supports these views. While the earliest known find of einkorn is from Jarmo, Helbaek believes that it was always a cereal of minor importance outside Turkey and the chief finds of prehistoric einkorn are from Turkey. Its presence in prehistoric and early dynastic Egypt has been definitely disproved.[33] Emmer, on the other hand, wherever described in early deposits is of a uniform and highly specialized type that cannot have derived from crops of the first generations of cultivated plants, "not even from the first hundred generations." Such developed emmer occurs abundantly in the predynastic stratification of Egypt, and throughout Palestine, Syria, Turkey, and Iraq. Emmer was cultivated along the Danube and spread throughout Neolithic Europe. I believe that it is possible that the tetraploid emmer was cultivated in a Palestinian hearth before diploid einkorn was domesticated.

Botanical considerations and prehistoric plant identifications would appear to support the view suggested earlier that the beginning of cereal cultivation lies in the Natufian cultures. Natufian "Mesolithic" sickles indicate that domestic varieties, i.e. with a tenacious rachis, had appeared by the end of the twelfth millennium B.C. at the latest. Sickles are not suited for harvests of grains whose axis and rachis shatter, scattering the kernels. That the Natufians harvested grasses for their grains and not merely for their stalks (e.g. for thatch) is indicated by their profuse production of grind stones and grinders. The massive pestles and mortars that have been found suggest that their grains were "glumed" or hulled like emmer and einkorn to extract the naked grain. It is also likely that

[31] *Ibid.*, p. 1079.
[32] On the characteristics of wheat species and their distribution and antiquity see Paul C. Mangelsdorf, "Wheat," reprinted from the *Scientific American*, July 1953 (San Francisco: W. H. Freeman & Co.), pp. 4–11.
[33] Helbaek, "Ethnobotany," p. 107.

emmer was the first domesticated grain, for its primary habitat coincides with a core area of Natufian culture in northern Galilee. Wherever else it is found, it appears as an already fully domesticated grain in the earliest deposits.

It is possible that the area of incipient cultivation on the flanks of the Zagros, including the small village community of Jarmo, received a stimulus for local domestication, if not actual seed grain imports, from the hill country of the southern Levant. Finds of emmer there belong to a crop of "conspicuously mixed character" with some grains appearing similar to the wild and some more typical of cultivated emmer.[34] A conspicuously mixed harvest may suggest that Jarmo is very close to the area where the transition from wild to domestic emmer was accomplished. Under conditions of incipient domestication a variable crop with a fairly high proportion of grain similar to wild emmer might be expected. But obvious as such a conclusion may be, I believe it is probably wrong, for the botanical and archaeological evidence from the Levant also supports a different interpretation. What seems likely is that the mixed harvest of Jarmo is evidence of an attempt to force a domesticated plant into an area that was ecologically unfavorable to it. The variability in the crop would then stem from the incorporation of the Turkish–Iranian emmer race which had invaded the fields of introduced domesticated emmer. The latter itself would become more variable in a new habitat.[35]

The question of why the cereals should have been domesticated first in their primary habitats remains. Harlan and Zohary point out that where natural stands are as dense as a cultivated field there is hardly any necessity for cultivation.[36] It is possible that early tillage and planting stem from the attempt to extend the range of the wild plant. Such attempts would accelerate the rate of mutation and favor the early appearance of domestic varieties. The location of villages like Jarmo and Jericho, both peripheral to the range of Natufian and cognate cultures, support these considerations.

A large variety of wheats has been developed, most of them hybrids of either emmer or einkorn. The forty-two chromosome wheats (hexaploids) contain the bread wheats, which are known only in cultivation and are the most useful today. The hybrid hexaploids which occurred first in emmer type fields were themselves subsequently selected for cultivation as the new bread wheats. Club wheat, *T. compactum*, appears as cultivated wheat ca. 2000 B.C. in Iraq and by the first millennium B.C. it is important in the Near East. In central Europe it appears as an important grain by 3000 B.C.[37] Although the grain appeared sporadically

[34] *Ibid.*, p. 102.
[35] The data obtained by B. Lennart Johnson by electrophoresis of crude emmer protein extracts would be in accord with this interpretation. "Tetraploid Wheats: Seed Protein Electrophoretic Patterns of the Emmer and Timopheevi Groups," *Science*, 158, No. 3797 (6 October 1967), 131–32.
[36] "Distribution of Wild Wheats," p. 1079.
[37] Andrews, "Genetic Origin of Spelt," p. 20.

in the Near East from the third millennium B.C., it was successful only outside the range of early grain cultivation, especially in areas of summer rains. The most important hexaploid is the common bread wheat, *T. aestivum* (formerly *T. vulgare*), which originated somewhere in northeastern Turkey and adjacent areas of Iran and the USSR. It is a naked, easily threshed wheat, yet the stalk of its inflorescence is tough and does not shatter in harvesting.

In contrast to other hexaploids, spelt, *T. spelta,* is a glume wheat. Current genetic theories assume that it developed as a hybrid between a variety of emmer or wild emmer and *A. squarrosa* either in the Near East or southeastern Europe. A southeast European origin has come to seem increasingly unlikely with the discovery of spelt cultivation near Shahr Kord in the western Esfahan province of Iran. The Iranian spelt is identical to European spelt.[38] Spelt probably intruded into cultivated crops of emmer and einkorn and became a minor fractional component of wheat introduced into the Zagros highlands or perhaps into southeastern Europe where cooler climatic conditions favored the suppression of emmer and einkorn and the emergence of spelt as a major crop. Spelt appears in first millennium B.C. deposits in central Europe and was very important in early second millennium B.C. Italy.

Spelt is thus a secondary domesticate. In its widest application the term secondary domesticate refers to any plant domesticated later than the commonly accepted original stock of plants of a domesticating culture. It usually implies a spread of the domesticating culture from its hearth to new areas, although this is not always the case. The secondary domesticate may be derived from the flora of the new areas of settlement, or it may have been carried along as a minor hybrid fraction of a crop or as a weed. Actual migration need not occur to accomplish secondary domestications, for various kinds of culture contact might provide the possibility for the spread of domestication techniques. Teff, for example, was added in Abyssinia to the introduced wheat crops. Spelt, another secondary domesticate, was a minor hybrid fraction of wheat in its original hearth; it became increasingly important in the north and westward migration of grain growing cultures, and achieved dominance in suitable ecological areas. Rye migrated as a weed with wheat, until it became a domesticated plant in its own right.

There are terms which more closely define the attributed function of a secondary domestication. "Substitute domestication" refers to the substitution of a newly domesticated local plant for the introduced older domesticate, which is then given up completely. Presumably some millets and sorghums are substitute domesticates for wheat in parts of Africa and Asia. Secondary domesticates in some areas acted as "bridging domesticates" allowing an original domesticate to pass through an area for which it was ecologically unsuited and to reach an area in which it might

[38] Herman Kuckuck and Elizabeth Schiemann, "Über das Vorkommen von Spelz und Emmer (Trit. Spelta L. und Tr. dicoccum Schübl.) im Iran," *Zeitschrift für Pflanzenzüchtung* 38, No. 4 (December 1957), 383–96.

again thrive. Millets and sorghums, for example, allowed the bread-grain cultivating cultures to reach eastern Asia. Wheat was not abandoned, but its low yield in central and southwest Asia led to a heavy reliance on millets, whose importance declined as soon as areas where wheat cultivation could be pursued were reached.

MILLET AND SORGHUM. Millets, unlike wheat and barley, do not constitute a genus but include plants of different genera. In agricultural usage millets are small seeded cereals with abundant foliage and smooth, groove-less grains. The term is sometimes used to include sorghum. Common millet, *Panicum miliaceum*, and Italian foxtail millet, *Setaria italica*, may be Near Eastern domesticates. While the oldest archaeological evidence is Near Eastern, these millets are scarce in the deposits and only become important on the European and Asian periphery of the Near Eastern hearth.[39] Common millet was widely used in "Neolithic" Holland, the Ukraine, and Roman Italy. In parts of Europe this millet acted as a bridging domesticate, allowing the introduction of bread grains beyond the area of the millet. It did not reach England in the early period, but other grains did. Subsequently, of course, millet was displaced even within the areas where it was once dominant when ecologically suitable bread grains developed through hybridization or mutation. India was an important hearth of secondary domestication for millet and sorghums, e.g., durra, *Andropogon sorghum*, pearl millet, *Pennisetum glaucum* (or *P. spicatum*), and finger millet or raggee, *Eleusine coracana*. From India millets as well as sorghums reached China, where bread grains subsequently largely displaced millets.

Following Vavilov, a number of scholars have identified the center of domestication of some millets in China and others in Abyssinia. The sorghums and common millet, on this theory, were originally Abyssinian. H. Doggett, for example, believes that sorghums were first domesticated in Abyssinia by Caucasoid settlers who had brought wheat from the Near East and then domesticated sorghums in their new home, perhaps after eusorghums had invaded their wheat fields. In his view, domestic sorghums then spread into other African areas and into India at the end of the second millennium B.C. Sorghums would have reached China, then, both from India and from Africa.[40] According to another view, sorghums were first domesticated by Near Eastern cultivators. The wild eusorghums of Africa would then be the weeds that escaped cultivation and spread widely in the native flora. There seems less ground for locating the hearth of common millet domestication in Abyssinia since its geographic distribution then becomes unintelligible. Common millet is unknown in archaeological finds from the Nile Valley, yet it should have spread there from Abyssinia in order to reach Europe.[41] Abyssinia was not a hearth of primary domestication, as Vavilov would have it,

[39] Helbaek, "Paleoethnobotany," pp. 112–13.
[40] "The Development of Cultivated Sorghums," p. 65.
[41] Helbaek, "Paleoethnobotany," p. 113.

but a center of secondary and substitute domesticates, including a number of millets and sanga cattle. These domestications facilitated the expansion of agriculture into other parts of Africa.

An argument for a Near Eastern–Indian hearth for millet is the coincidence in the distribution of cattle and millet in Africa, which has been noted by Emil Werth. In Africa it is the zebu raising peoples who grow millet, and since zebu is of Indian or southwest Asian origin, millet presumably would have arrived with the culture that introduced the zebu.[42] As far as a Chinese origin for millet is concerned, there is no evidence for its primacy there, and the mere fact of a great number of varieties points rather to the character of China as a boundary sphere of the grains. A Chinese hearth has also been suggested on cultural grounds, for millet played an important role in Chinese ritual.[43] But the argument can with greater justice be reversed on cultural grounds, for millet was important in Chinese kingship rites which seem clearly to be derived from the sacral kingship rites of the Near East.

Nor does the persistence in China until recent times of early methods of the preparation of millets for food indicate that millets were of great antiquity there. The persistence of conservative usages is particularly common in areas of relatively late introduction. This often applies to dietary habits as much as to other social and cultural traits, and the persistence of old cooking methods that have been superseded in other areas cannot, in itself, be interpreted to mean that either the methods or what is being cooked are original to the area. Almost all grains were prepared as gruel before the invention of baking. In Europe millets were generally given up with the spread of the art of flour baking. Suitable bread grains, which, before baking, had been used like millets, in the preparation of gruels, displaced millets as field crops. However, the older ways of preparing grain for food persisted, chiefly in ritual usage, e.g., Biblical *soleth* and the barley gruel, *kikeon*, of ancient Greece. In China gruels remained a favored method of food preparation and millets therefore remained more important than in Europe and west Asia. But even China succumbed to the introduction of other grains and methods of food preparation and probably gave up the cultivation of some millets. It is possible that *Panicum spontaneum* is a weedy descendant of an old cultivated form of millet. Similarly crab grass, *Digitaria sanguinale*, and barnyard millet, *Echinochloa crus-galli*, have elsewhere in Eurasia descended to weed level from cultivation, although one form of it, Japanese barnyard millet, *E. frumentacea* (*Sanwa* in India), remains important in east Asia.

RICE. The migration of grain farming peoples eastward into India and Southeast Asia led to another secondary domestication of far-reaching importance—that of rice. Agriculturally rice might be classified as a

[42] Werth, *Grabstock*, pp. 21, 98.
[43] Marcel Granet, *Chinese Civilization*, trans. Kathleen E. Innes and Mabel R. Brailsford (New York: Meridian Books, Inc., 1960), pp. 144–45.

millet, and its dietary usages more or less parallel those of the important millets. The many varieties of rice are grouped into two categories: mountain, *Oryza montana,* and lowland rice, *O. sativa.* The former probably indicates the type of terrain of original domestication. In the course of descent into the wet monsoon tropics, a wild variety of mountain rice, *O. fatua,* presumably entered into cultivated areas, achieving ecological dominance in wet lowland soils. From India rice was probably introduced into China, Korea, and Japan, as well as into the islands of the Indonesian archipelago and the Philippines.

The routes of dispersal of cultivated rice, like those of other Old World grains of worldwide economic significance, have not been completely traced. Imprints of rice have been identified on Yang-shao pottery of ca. 3000 B.C., and a Chinese character for "rice" occurs in Shang dynasty writings of ca. 1000 B.C. In many of the Chinese literary records between the first and sixth centuries A.D. rice is treated as a weed appearing in cultivated lands in southern China. It was apparently considered both a nuisance and a famine food. It is difficult to decide whether these records refer to a wild progenitor or to a plant that descended to a weed state after earlier cultivation. Nor is the problem solved by Ting Yin's successful breeding of new rice varieties from "wild" varieties found at Le-Fu mountain in Kwangtung and in the vicinity of Canton.[44] In all probability, the traditional theory of the dispersal of cultivated rice from India to China, Korea, and Japan, and from the same hearth to the Indonesian archipelago and the Philippines, is correct. Mutants probably developed in the course of its movement from southern China, and then dispersed, escaping from cultivation. Westward, rice spread to Sumer and to northern Mesopotamia in the last half of the first millennium B.C., but to Europe and America only in the fifteenth and seventeenth centuries A.D., respectively.[45]

RYE AND OATS.   The domestication of rye and oats is of interest as an illustration of the process of a weed that was dispersed with a host crop being transformed into an important domestic grain. Both rye and oats are substitute domesticates whose domestication was achieved in the early first millennium outside the Near East. Rye, *Secale cereale,* appeared as a weed in Britain and Denmark, inadvertently introduced by Iron Age immigrants, and it quickly became dominant not only there but in many parts of northern Europe. Its first known cultivated form was found in the eastern Alps, and it became a regularly cultivated crop in most European areas under Roman influence. There is debate concerning which of the wild varieties of rye is the progenitor of domestic rye, since many appear to be weedy races that escaped from cultivation. It

[44] This material is based largely on Tsu-kuei Chen, *Rice Cultivation in Chinese Documents,* Agricultural History Research Series, Vol. II, Nanking: Institute of Agriculture, Academia Sinica (Nanking: Science Press, 1960) which was translated for me by Tsung-lu Kao. For further relevant material see Ping-Ti-Ho's outstanding study "Early-Ripening Rice in Chinese History," *Economic History Review,* IX, No. 2 (1960), 200–18.

[45] Ludwig Reinhardt, *Kulturgeschichte der Nutzpflanzen* (München: Verlag von Ernst Reinhardt, 1911), I, 53–57.

was probably either *Secale cereale ancestrale* or *S. cereale afghanicum*, both of which are found between Afghanistan and Turkey. In any case, rye invaded wheat and its success in this is still attested by the fear which Near Eastern farmers have of invasion of their wheat fields by weedy rye. In the course of its spread as an invader of wheat, rye developed a tough axis, thus being domesticated with the wheat in which it stood, and can therefore be said to have domesticated itself. Rye, like wheat, contains gluten and can be made into a porous bread. This facilitated the farmer's adoption of this "self-domesticate." [46]

The most important cultivated oat, *Avena fatua sativa*, derives from a wild hexaploid plant, *A. fatua*, which invaded wheat fields in the Near East and in eastern Europe.[47] This weed was introduced into Europe along with spelt and emmer during the latter part of the first millennium B.C. When it was first actually cultivated in Europe is uncertain; it seems clear that it was never cultivated in the Near East. Greco-Roman authors usually mention oats as a fodderplant and as a famine food. It became a major staple (oat meal) in medieval northern and northwestern Europe.

The ability of oats to supplant wheat was facilitated, some believe, by the deterioration of climate in the post-Atlantic period. According to this view, the agricultural colonization of Europe went forward under climatic conditions warmer and more humid than at present in the period from the sixth to the middle of the second millennium B.C. (the Atlantic phase). The early Near Eastern colonist in Europe encountered conditions which, in terms of agricultural ecology, were not significantly different from conditions in the areas of Near Eastern agriculture.[48] Subsequent climatic deterioration led to the decline of wheat and barley, and to the ascendancy of rye and oats. One oat, *A. fatua*, was particularly favored by the increasing length of cold winters, for its seed is not viable if winters are not cold.[49] Oats, like rye, also made it possible for settlers to colonize areas too cold or too high for the older Near Eastern grains.

The questions of why the plants discussed were the first to be domesticated and why they have remained the chief food staples are intriguing. The variety of plants used by early man for food was very great: why were the grains the earliest domesticates? Perhaps one reason is that the relationship of carbohydrates and proteins in these grains is such that in case of need, man can indeed live by bread alone.[50] Also, grain grows in dense natural stands which would have proved attractive to food gatherers. Increased rates of collecting and harvesting as a result of denser human settlement in the area would lead to increased variability and extragenic mutation of plants, whether as a result of selective harvesting, intentional or inadvertent dispersal of harvested grains, plant colonization of human refuse heaps, or some other processes incidental

[46] La Baume, *Frühgeschichte*, pp. 30–31. Helbaek, *Paleoethnobotany*, pp. 113–12.

[47] G. D. H. Bell, "The Comparative Phytogeny of the Temperate Cereals," Chapter IV in Hutchinson, *Crop Plant Evolution*, pp. 89–96.

[48] Butzer, *Environment and Archaeology*, pp. 447–49.

[49] Bertsch, *Kulturpflanzen*, p. 79.

[50] Rau, *Nutzpflanzen*, p. 221.

to the presence of man. The sickle harvesting of grain, for example, would scatter the seed of plants whose fragile rachis was likely to shatter, and the ground would be seeded more densely than under conditions of natural seed dispersal. The increased density of sprouting seed might have conferred differential competitive advantages on certain strains. Another reason why the grains may have been particularly attractive is because the wild forms are sometimes perennial, and in any event, harvesting did not diminish the productivity of an area, but rather increased it, allowing for a second crop in one season.

The reduction of life span between germination and ripening which occurred in the domestication of grains aided in their spread toward the tropics and into northern latitudes. In ecologically difficult areas short-lived plants have competitive advantages over longer lived plants. More slowly ripening varieties remain for a longer time in danger of pest damage, plant infection, and bad weather. What facilitated the reduction in life span was perhaps inadvertent selection of quickly flowering and ripening grains which would then become the greater part of the harvest. In moving from one area to another it would again be the quickly flowering fraction that would have the competitive edge. Another feature of Near Eastern grains that encouraged spread is their derivation from plants adapted to survival in dry summers through dormancy, and their ability to grow actively at low temperatures and in weak daylight during rainy winters.[51] Once carried to northern areas, provided the winter was not too cold, these characteristics enabled the grains to do better than indigenous related plants. Neither the shift to wet summer growing seasons nor survival of seed in snow covered soil posed major problems to north and westward Near Eastern or Mediterranean migrating cultures.

LARGE-SEEDED LEGUMES.   Historically less important are the large seeded legumes. Peas, *Pisum sativum*, lentils, *Lens esculenta*, and some kinds of vetchling have been recovered at Jarmo and in fourth millennium B.C. Egyptian sites. Chickpeas, *Cicer arietinum*, of some importance today in southern Europe and western and central Asia, are found in Palestinian deposits of the fourth millennium B.C. The ancestral wild pea probably derives from *P. elatius*, whose range extends from the Mediterranean to Afghanistan. It is generally believed that the pea was a secondary domesticate, starting its domestic career as a weed invading grain fields. The very wide distribution of wild peas may have facilitated repeated local secondary domestications in Iran, Afghanistan, and Pakistan as well as in the Balkans, Turkey, and Switzerland. Since pea seeds are free of the toxic or bad tasting substances found in most legumes, peas may have not been considered weeds by farmers, but may have been welcomed. Lentils, it has recently been determined, are derived from a Near Eastern wild lentil, *L. schmittspahni*, and not, as was for-

[51] Cf. J. P. Cooper, "The Evolution of Forage Grasses and Legumes," Chapter VII in Hutchinson, *Crop Plant Evolution*, pp. 155–59.

merly thought, from a central Asian plant.[52] The broad bean, *Vicia faba,* has been cultivated in the Near East at least since the third millennium B.C. Wild broad beans are found on the Mediterranean coast from Algeria to Turkey.[53]

FRUITS. Fruits of trees and shrubs were of course used long before any attempt was made to propagate them systematically. Indeed many of the berry bearing plants and fruit trees have only been cultivated in recent times, and other fruit supplying shrubs and trees are still wild. The obvious incentive for the domestication of berries was the discovery of fermentation as a method of food preparation, a discovery first associated with the preparation of grains. Malting occurs naturally when farinaceous grain germinates, converting its starch into maltose and dextrose. Malting was used to make seeds and the hulled fruit of grain edible. Long after the invention of baking in Mesopotamia, about forty percent of cereal production was still used for brewing. The importance of fermentation can be seen in Sumerian texts which mention eight barley beers, eight emmer, and three mixed beers.

The effects of fermentation must have been known in the Upper Paleolithic, but most collected wild fruits contained too little sugar to be suitable as base material.[54] Grains were combined with honey (ambrosia, mead, and nectar were some of the earliest fermented drinks). In any case in Mesopotamia and Egypt, some time after the introduction of the fermentation of grain and honey, the fermentation of grapes and dates became common. A range of fermented fruit drinks gradually displaced the numerous Near Eastern beers.

The history of "royal" wines suggests that domestication of some grapes may have been undertaken as a ceremonial task by Near Eastern kings. The regular cultivation of vines can be traced in archaeological deposits to the fourth millennium B.C. Wild grapes, *Vitis silvestris,* grew widely in Europe and Asia but the original domestication of wine grapes, *V. vinifera,* seems to have taken place between the Caucasus, eastern Turkey, and the Zagros ranges. Grapes occur in the late fifth millennium B.C. deposits. It is certain that viticulture was practiced and wine was made in Mesopotamia and brought to Egypt before 3000 B.C. The king Urukagina (ca. 2350 B.C.) and Gudea, viceroy of Lagash (ca. 2100 B.C.) in southern Mesopotamia had large vineyards and by the time of Abraham, vineyards with 29,000 vines are mentioned near Haran in northern Mesopotamia. The importance of wine was such that Assyrian kings

[52] Walther Rytz, "Entstehung und Herkunft der Kulturpflanzen," Chapter V, or *Urgeschichte der Schweiz,* ed. Otto Tschumi (Frauenfeld: Verlag Huber & Co., 1949) I, 108.

[53] On this bean and its ancient mythologocial and ritual aspects as well as the light it throws on ethnic origins see Alfred C. Andrews, "The Bean and Indo-European Totemism," *American Anthropologist,* 51, No. 2 (April-June, 1949), 274–92. Eugene Giles, "Favism, Sex-Linkage, and the Indo-European Kinship System," *Southwestern Journal of Anthropology,* 18, No. 3 (Autumn, 1962), 286–90.

[54] R. J. Forbes, *Studies in Ancient Technology,* 2nd. ed. (Leiden: E. J. Brill, 1965), p. 62.

considered it of strategic importance to cut off an enemy's wine supply.[55]

The library of Asshur-ban-apal (668–626 B.C.) yielded a list of the ten best wines headed by the "wine of Izalla" followed by the wine of Helbon (cf. Ezekiel 27:18). Asshur-nasir-apal II (ca. 884–859 B.C.) and Sanherib (704–681 B.C.) pursued a systematic search for exotic vines which they tried to acclimatize in northern Syria and Mesopotamia. There is no doubt that the original center of viticulture was in the Near East and that the art spread from there with grain agriculture into Mediterranean lands. The vine is not indigenous to Egypt and although it occurs in predynastic times, must be regarded as an imported domesticated plant. Classical authors do not approve of Egyptian wines. Martial thinks that vinegar tastes better, and Strabo recommends that Egyptian wine be mixed with seawater to avoid headaches. In Palestine viticulture was only slightly less important than olive and fig cultivation.[56]

Arboriculture, like viticulture, had its hearth in the Near East. Olives and dates are found in Palestine and Egypt of the fourth millennium B.C., but both may be assumed to have been domesticated earlier. Important fruit trees mentioned in Sumer include carob, fig, date, apple, apricot, pear, quince, medlar, and peach. While apple, pear, cherry, fig, olive, and wine grapes were gathered as wild fruits elsewhere, notably in Europe, the concept of arboriculture was introduced from the Near East together with new domesticated varieties.[57] The wild forms of sweet cherries, *Prunus avium*, are widespread, but the earliest domestic record is Near Eastern. Other varieties are found wild only in the Caucasus and in Turkey (morelle cherries) or in the Mediterranean basin (e.g. maraschino). The plum, *P. domestica*, is hybrid between cherry plum, *P. cerasifera*, and blackthorn, *P. spinosa*. Schwanitz believes that the region of origin was in the contact zone of the two species in the Near East.[58] The peach and the apricot may have been originally domesticated in China and are assumed to have spread westward to the Near East and Europe. Peaches reached Persia ca. 130 B.C. and apricots appeared in the Near East in the same period.

A number of berries, including blackberries, gooseberries, raspberries, and red and black currants were not cultivated until between the sixteenth and eighteenth century A.D. The common wild strawberry, *Fragaria vesca*, was first cultivated in fourteenth century France, but extensive cultivation dates to the introduction of the scarlet Virginian and Chilean strawberries into Europe in the seventeenth and eighteenth centuries, respectively. The hybridization of the two in Holland produced the pineapple strawberry, *F. grandiflora*, which is ancestral to most modern commercial varieties.

PLANT FIBRES.   Plants were used from the earliest times for purposes other than food: for shelter, clothing, and drugs, especially. Many of these

[55] *Ibid.*, p. 73.
[56] *Ibid.*, p. 78.
[57] Helbaek, "Domestication of Food Plants," p. 371.
[58] *Cultivated Plants*, p. 153.

plants were never domesticated, though the use of many of them, for example as poisons and medicines, has been continuous. There is new interest in some, so that efforts to domesticate them in order to increase the yield of the active agent are probable.[59]

Judging by the archaeological record, the earliest major fibre plant to be domesticated was flax, *Linum usitatissimum*, whose wild progenitor, *L. angustifolium*, differs from the domestic form only in its smaller and more open capsule. Large seed and closed capsules are apparently the result of the forced existence of the species beyond its natural habitat. Wild flax is found in the Near East and all around the Mediterranean from the Black Sea to the Canaries, and seed imprints are found in the Halaf period of the fifth millennium B.C. in Kurdish Iraq and further south in the alluvial plains. It is known in the Levant at the same time. In Egypt domesticated flax occurs in the early Fayum Neolithic, but no wild form has been found. Herodotus seems to imply that Egypt obtained flax from the Caucasus in saying that "These two nations (Egyptians and Colchians) weave their linen in exactly the same way, and this is a way entirely unknown to the rest of the World. . . ."[60] But the geographer Strabo is wary of such historiography: "Their (the Colchians') linen industry has been famed far and wide, for they used to export linen to outside places; and some writers, wishing to show forth a kinship between the Colchians and the Egyptians confirm their belief by this."[61] According to Forbes, "We just do not know what part of Western Asia was responsible for the discovery and working of the fibres of flax."[62] We do not even know if the greater incentive in domestication was the fibre or the oil of the plant (linseed), although the presumption is strong that the seed capsules were used as food long before fibre production.

In part the significance of flax was religious: the use of linen often went hand-in-hand with the rejection of wool. Wool was regarded as ritually unclean in ancient Egypt, an attitude that was not confined to Egypt. Philostratus writes,

In what way is linen better than wool? Was the latter not taken from the back of the gentlest of animals, from a creature beloved by the gods who do not disdain themselves to be shepherds and, by Zeus, once held the fleece to be worthy of a golden form. . . . On the other hand linen is grown and sown everywhere, and there is no talk of gold in connection with it. Nevertheless, because it is not plucked from the back of a living animal, the Indians regard it as pure, and so do the Egyptians, and I myself and Pythagoras on this account have adopted it as our garb, when we are discoursing or praying or offering sacrifice. . . .[63]

---

[59] Cf. Margaret B. Kreig, *Green Medicine* (Chicago: Rand McNally & Co., 1966).

[60] Herodotus II, 105. Quoted after George Rawlinson, trans., and Manuel Komroff, ed., *The History of Herodotus* (New York: Tudor Publishing Co., 1956), p. 115.

[61] Quoted after R. J. Forbes, *Studies in Ancient Technology* (Leiden: E. J. Brill, 1964), IV, 27.

[62] *Ibid.,* p. 27.

[63] *Ibid.,* p. 28.

The Biblical prohibition against mixing linen and wool in one fabric (Lev. 19:19) is of course still followed by orthodox Jews today, and is known as the "sha'atnez" prohibition.

COTTON. The cultivated species of old world cottons, *Gossypium herbaceum* and *G. arboreum*, are recent secondary domesticates. The cultivated cottons are predominantly quickly ripening annual plants, whose short-term growing habit was achieved through domestication. Primitive forms of the cultivated species, like wild cottons, are perennial, and such cottons are still widely spread in India, Southeast Asia, Africa, and the New World. New World domestic cottons are tetraploid, one set of chromosomes homologous with the chromosome complement of Old World (diploid) cottons and one set with the long chromosome complement of New World wild species. The New World tetraploid cottons have spread widely into the Old World, chiefly after the age of discovery.

The history of cotton has been particularly hard to unravel: the plant establishes itself readily in waste places and as an invader on cultivated land, so that it is difficult to determine whether wild plants that are discovered are the ancestral form of domestic cottons or "feral" plants that escaped cultivation from field-planted cottons. It is possible that cotton first entered domestication as a wild invader of fields, and perhaps started as a substitute domesticate for flax. The original cultivators of cotton seem to have been the Indians. The perennial shrub is sacred in India and was cultivated by the early Indus river civilization. Cotton was exported to Mesopotamia before 1500 B.C., and Sanherib (704–681 B.C.) introduced tree cotton into Assyria. The planned introduction of Indian cultivated tree cotton, *G. arboreum indicum*, into Mesopotamia may have initiated the changes that made the plant a truly domestic one. In southern Mesopotamia and the Persian Gulf area cotton did well and was grown from the eighth century B.C. The perennial Indian cottons are also ancestral to the domestic cottons of Indo-Malaya and Africa. Their spread eastward and the development of annual varieties of arboreum was slow. Cotton goods were imported from India and Java in the fifth century A.D., but cotton was not cultivated in China before the eleventh century A.D. The Chinese must have been among the first to develop annual varieties of arboreum since most of the crop is grown in cold-winter regions. The Chinese *G. arboreum sinense* is the earliest fruiting form of cotton.

While arboreum eventually developed annual forms, the earliest annual forms derive from *G. herbaceum*, primitive cultivated forms of which persist in southern Arabia and at the head of the Persian Gulf. These may have been brought in the same "royal" domestication which introduced *G. arboreum* into Mesopotamia. Short-season annual varieties of *herbaceum* were introduced into Persia and central Asia, areas which were linked culturally and politically for long periods with Mesopotamia. The first annual cottons in Indian agriculture are derived from *G. her-*

baceum persicum and were not introduced into India until the eighteenth century A.D.[64]

The New World

The relation of New World to Old World plant domestication is unsolved. Cotton presents particularly intriguing problems, for cultivated tetraploid Gossypium species may derive from a prehistoric combination of Old and New World wild cottons. Cotton fabrics found in early levels at Huaca Prieta, Peru, date at least to 2500 B.C., thus coinciding in age with the oldest recorded cottons in the Old World. Mexican finds are older, going back in Tehuacan to around 3500 B.C. Perhaps Old World cottons were introduced to the New World and subsequently resulting hybrids were returned to the Old World. George F. Carter considered Huaca Prieta cotton an example illustrating ancient transoceanic contact.[65] However, the possibility of natural hybridization has not thus far been definitely discarded. The problem is further complicated by the fact that the domesticated cottons of Middle and South America are distinct. Recent studies show that neither hypothetical widespread ancestors nor recent origin through crossing with Old World cottons can explain the origin of New World cottons.

The case for an independent origin of domestication relies on the different assortment of plants and the relative importance of different kinds of domesticates in the agricultural complexes of the Old and New World. A heavy reliance on many different cereals in the Old World contrasts with the importance of maize in the New. The abundance of legumes in the Americas contrasts with the relative scarcity of leguminous plants in the Old World. Similarly, the relatively greater importance of *Cucurbita*, different kinds of root crops, and relish plants in the New World, and the differences between the Old and New World use of similar species all provide a fairly sharp contrast between the two agricultures. Furthermore, the New World did not domesticate plants particularly suited for sowing broadcast, reaping and threshing, nor were domesticated animals incorporated into the agricultural systems for plowing, manure, or milk production.[66] Of course, such comparisons are in themselves of limited value, for agricultural systems in the Old World which are known to be related historically show similar wide variation. The case

[64] Lyle L. Phillips, "The Cytogenics of *Gossypium* and the Origin of New World Cottons," *Evolution*, 17, No. 4 (December 1963), 460–69, and P. A. Fryxell, "Stages in the Evolution of Gossypium L.," *Advancing Frontiers of Plant Sciences*, 10 (1965), 31–56. In addition the material on cotton is based on Sir Joseph Hutchinson, "The History and Relationships of the World's Cottons," *Endeavour*, xxi, No. 81 (January 1962), 5–15, and on Forbes, *Ancient Technology*, IV, 43–49.

[65] "Movement of People and Ideas across the Pacific" in *Plants and the Migrations of Pacific Peoples: A Symposium*, Tenth Pacific Science Congress, Honolulu, Hawaii, 1961, Jacques Barrau, ed. (Honolulu: Bishop Museum Press, 1963), pp. 7–22.

[66] Herbert G. Baker, *Plants and Civilization* (Belmont, California: Wadsworth Publishing Company, Inc., 1965), 45–6.

for an independent New World domestication has been strengthened by spectacular archaeological finds, particularly in southern Tamaulipas and the Tehuacan valley of southern Puebla in Mexico. "Incipient agriculture" has been identified there at the beginning of the seventh millennium B.C.[67]

However the case against an independent origin cannot be considered closed. There is no doubt that American Upper Paleolithic/Mesolithic cultures show striking affinities with older Eurasian cultures.[68] That Old World migrants were the "incipient" domesticators of the New World is plausible.

A suggestion for an Old World stimulus for American incipient agriculture comes from the bottle gourd *Lagenaria siceraria*. This gourd, perhaps the oldest dated plant domesticate in the New World, is of great antiquity in both the Old World and the New. In the Old World it was used by Mesolithic folk for ritual as much as utilitarian purposes (e.g., ceremonial rattles, fishing floats, jars). While *Lagenaria* may have reached the New World accidentally by drifting on ocean currents,[69] it is so widespread in preagricultural and early preceramic sites from Chile to the United States that it is difficult to account for its distribution unless it was carried by man. There is no evidence that the wild gourd was a widespread self-propagating plant in the New World.[70]

Mainland Middle America is considered the most important center of American plant domestication. Pepo or summer squash and pumpkins, *Cucurbita pepo*, may have been domestic by 7000 B.C. in Tamaulipas; cushaw and walnut squash, *c. moschata* and *C. mixta*, are very early in southern Mexico.[71] Maize or Indian corn, *Zea mays*, is probably the most important contribution of aboriginal American agriculture to the world's food crops. Its origin and the process of its domestication, however, are unsolved problems. Maize does not occur in the oldest levels. The oldest maize found in archaeological levels is dated between 5200 and 3000 B.C. However, no wild maize is found in the Americas that might

[67] Richard S. MacNeish, "The Origins of New World Civilization," *Scientific American* 211, No. 5 (November 1964), 29–37. For further references see Richard S. MacNeish, "The Food-Gathering and Incipient Agriculture Stage of Prehistoric Middle America," in *Handbook of Middle American Indians*, Robert C. West, ed. (Austin, Texas: University of Texas Press, 1964), I, 413–26. A list of cultivated plants of Middle America is given by Paul C. Mangelsdorf, Richard S. MacNeish and Gordon R. Willey in "Origins of Agriculture in Middle America," *ibid.*, 434–35. On the origins of American agriculture see also Charles B. Heiser Jr. "Cultivated Plants and Cultural Diffusions in Nuclear America," *American Anthropologist*, 67, No. 4 (August 1965), 930–45; C. Earle Smith, Jr., "The New World Centers of Origin of Cultivated Plants and the Archaeological Evidence," *Economic Botany*, 22, No. 8 (July–September 1968), 253–66; and Hugh C. Cutler, "Origins of Agriculture in the Americas," *Latin American Research Review* III, No. 4 (Fall 1968), 3–21.

[68] Hansjürgen Müller-Beck, "Palaeohunters in America: Origins and Diffusion," *Science*, 152, No. 3726 (27 May 1966), 1191–1210. Cf. William J. Mayer-Oakes, "Early Man in the Andes," *Scientific American*, 208, No. 5 (May 1963), 126–28.

[69] T. W. Whitaker and G. F. Carter, "A Note on the Longevity of Seed of *Lagenaria siceraria* (Mol.) Standley after Floating in Sea Water," *Torrey Botanical Club Bulletin* 88, 104–6.

[70] Cutler, "Origins of Agriculture in the Americas," p. 14.

[71] Hugh C. Cutler and Thomas W. Whitaker, "History and Distribution of Cultivated Cucurbits in the Americas," *American Antiquity*, 26, No. 4 (April 1961), 469–86.

be designated the parent of the domestic varieties. A number of related plants exist, including teosinte (the Aztec's "teocintle"), and *Tripsacum*, a native American grass. Teosinte, which crosses freely with maize, has been variously identified as an ancestor or a late hybrid of maize and *Tripsacum*. *Tripsacum*, supposed to be especially close to maize, is further removed from it than many Asian and African grasses, including African sorghums and Asian coix.[72] In any event, teosinte appears as a natural element in the vegetation and in Mexican archaeological sites, supporting the idea that maize was domesticated there. Its domestication may have followed that of another grain, foxtail millet, a species of *Setaria*, which was apparently domesticated and then given up.[73]

Other important Middle American plant domesticates include lima bean (*Phaseolus lunatus*), runner bean (*Phaseolus coccineus*), tepary bean (*Phaseolus acutifolius*), manioc (*manihot esculenta*), papaya (*Carica Papaya*), tomato (*Lycopersicon esculentum*), various species of red pepper (*Capsicum*), and of cacao (*Theobroma* spp).

These Middle American domesticates were transmitted to South America where new varieties and related species were subsequently domesticated. Conversely South American domesticates reached the agricultural area of Middle and North America. Among them were tobaccoes (*Nicotiana tabacum* and *N. rustica*), peanut (*Arachis hypogaea*), guava (*Psidium guajava*), potato (*Solanum tuberosum*), and various jack beans (*Canavalia* spp). From the archaeological and botanical information available, it appears then that Middle America (particularly Mexico) and South America (notably Peru) are two focii of plant cultivation in the New World. Agriculture in the intervening areas is more recent.[74]

Of the two agricultural focii, Middle America is the older. But apart from the unresolved question of origin, the relation between Middle and South American agriculture poses difficult questions. The center of differentiation of the peanut is in Brazil, but the oldest finds are Peruvian (ca. 750 B.C.).[75] The center of wild tomatoes is in northwestern Peru, but the area of greatest domestic varietal diversity is in Mexico.[76] Whether Brazil, Mexico, or Peru should be considered as domestication hearths of these plants or centers of speciation at a prehistoric limit of dispersal depends on future botanical and archaeological work.

The view that has been proposed here is that the Near Eastern center of plant domestication is the primary Old World hearth, and that the domestication of plants in the Old World at least was basically a single

[72] Major M. Goodman, *The History and Origin of Maize*, North Carolina Agricultural Experiment Station, Tech. Bul. No. 170 (October 1965), 5–7. It seems clear, however, that maize was in the Old World in pre-Columbian times. Cf. Robert Heine-Geldern, "Kulturpflanzengeographie und das Problem vorkolumbischer Kulturbeziehungen zwischen Alter und Neuer Welt *Anthropos* 53 (1958), 361–402.

[73] E. O. Callen, "The First New World Cereal," *American Antiquity* 32, No. 4 (October 1967), 535–38.

[74] Cutler, "Origins of Agriculture in the Americas," pp. 9–10.

[75] Margaret A. Towle, *The Ethnobotany of PreColumbian Peru*, Viking Fund Publications in Anthropology No. 30 (New York: Wenner-Gren Foundation for Anthropological Research Inc., 1961), p. 108.

[76] Cutler, "Origins of Agriculture in the Americas," p. 11.

invention that was made in the area of what is now Israel, Jordan, Lebanon, Syria, Turkey, Iraq, and Iran. From that area plant domestication spread to peripheral areas. Many plants were first domesticated in these secondary centers, including central and western Europe, or in tropical centers in India, Southeast Asia, and Africa. The secondary centers were often productive in developing new varieties of old domesticates as well as in incorporating hitherto wild plants into cultivation. But what is important is that these areas are contiguous to the original domesticating hearth and that not only the concept of domestication, but basic techniques of cultivation reached them from the primary core. Near Eastern crops were transferred in the process, but were often abandoned at the limits of their ecological or economic range.

Most students are currently in agreement that domestication in the New World was an independent invention, although made at a later period. The case for Old World stimulus is not closed; there are hints of such a possibility, but the case cannot be well supported at this time.

CHAPTER 5 *the domestication*
of animals

In the case of animals, the first question becomes, What is the natural range of the wild ancestors of the four major food animals first known to be domesticated: goat, sheep, cattle, and pig? While the natural range of each of these animals is different, as the accompanying map shows, all of their ranges overlap in western Asia. It is possible that each of them was domesticated there. At any event it must be assumed that domestication occurred somewhere in or near the range of the wild ancestor, where ecological conditions were suitable for the joining of man and prodomestic animals in the domestication enterprise. This necessity for the conjunction of man and animal led to the riverine-oasis theories or propinquity hypothesis described earlier, and is basic to various speculations about domestication as rooted in human and animal sociability, territorial behavior, and herd instincts.[1]

[1] For an extensive recent coverage of the history of domesticated animals see F. E. Zeuner, *A History of Domesticated Animals* (New York and Evanston, Ill.: Harper & Row, Publishers, 1963). For summary statements of the problems raised by the domestication of animals and methods brought to bear on them see Wolf Herre, "Die züchtungsbiologische Bedeutung neuer Erkentnisse über Abstammung und Frühentwicklung von Haustieren," *Züchtungskunde*, 28, No. 5 (May 1956), 219–29, and Herre's "Der heutige Stand der Domestikationsforschung," *Naturwissenschaftliche Rundschau*, 12, No. 3 (March 1959), 87–94. Berthold Klatt, *Haustier und Mensch* (Hamburg: Richard Hermes Verlag, 1948). N. Nachtsheim, *Vom Wildtier zum Haustier*, 2nd. ed. (Berlin: Paul Parey, 1949). An indispensable guide to the often contradictory terminology of domestic animals is I. L. Mason, *A World Dictionary of Breed Types and Varieties of Livestock*, technical communication No. 8 of the Commonwealth Bureau of Animal Breeding and Genetics (Farnham Royal, Slough, Bucks.: Commonwealth Agricultural Bureaux, 1951). On animal classification authoritative works are J. R. Ellerman and T. C. S. Morrison-Scott, *Checklist of Palaearctic and Indian Mammals, 1758–1946* (London: Print. Order of the Trustees, British Museum, 1951), and Ernest P. Walker, *et al.*, *Mammals of the World* (Baltimore, Md.: The Johns Hopkins Press, 1964), II.

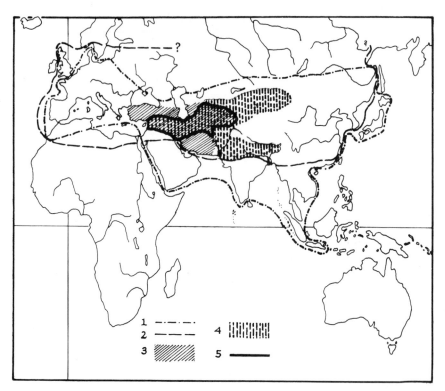

Figure 7. *Distribution of Wild Ancestors of Major Domestic Old World Animals. 1, Pig; 2, Cattle; 3, Goat; 4, Sheep; 5, Region of range overlap of sheep, goats, pigs, and cattle.*

THE GOAT.   The goat, *Capra hircus*, perhaps the first domestic herd animal, is generally thought to derive from the bezoar or pasang goat, *Capra hircus aegagrus*,[2] the only wild goat whose range extends from the Indus Valley westward through the Near East. Formerly this goat's range extended into Greece and perhaps as far north as the Austrian Alps, and on Crete today some more or less pure examples of the wild ancestral strain remain. There is no evidence that the ibex *Capra ibex* was ever domesticated.

The changes that distinguish domestic from wild goats are exceedingly difficult to trace osteologically. This difficulty has led Charles A. Reed to state his "complete disbelief in the validity of most of the published identifications used in analysing origin and spread of prehistoric domestic goats and sheep."[3] Not only is it difficult to tell domestic from

[2] D. R. Harris, "The Distribution and Ancestry of the Domestic Goat," *Proceedings of the Linnean Society of London,* 173 Session, 1960–61, Pt. 2 (April 1962), 79–91.
[3] Reed, "Archaeological Evidence," p. 129.

nondomestic goats, but sheep and goats can hardly be told from one another, except for parts of the skull, horn cores, phalanges, and metapodials, where these occur in the bone finds. Even the use of key measurements of bones held to be characteristic cannot identify the prodomestic animal. Zeuner and Reed believe that horns are the only characteristic upon which identification of domestic goats can be made in animal remains. The modern wild goat assumed to be descended from the ancestor of the domestic goat has scimitar-shaped horns, quadrangular in section (there are no modern wild goats with screw horns). All that survives of wild goat horns in prehistoric finds is the bony core which has a sharply keeled front and is also quadrangular in section.[4] Reed holds that lozenge-shaped sections of flattened medial surfaces of the cores are presumptive evidence of domestication. He believes the earliest domesticated goats in southwestern Asia were scimitar-horned, but that by the bronze age they were generally replaced by the screw-horned animals found today. The screw-horned animals can be recognized archaeologically by their twisted horn cores.[5] Reed dismisses claims by some European students to have found remains of wild screw-horned goats of "mid-European Pleistocene" date which they believe to be ancestral to *Capra hircus*. He believes these animals are domestic and of much later date.[6]

The first goats to be identified as domestic on the basis of their horns come from Jericho and were identified by Zeuner. Reed finds in the Jarmo deposits a record of the transition from wild to domestic goats, the latter having horn shapes similar to modern domestic goats. Apart from the possible but osteologically unverified presence of domestic goats in Natufian deposits of El-Khiam, the Jericho and Jarmo finds are the earliest so far.[7] At any rate domestic features had thus been acquired before the middle of the seventh millennium B.C. It is interesting that the chief change by which domestic goats can be identified seems to be of no economic importance. If screw-horns occurred as a mutation in the wild animal, they apparently did not survive. In the domestic state either the scimitar-horned animals were weeded out by intention or horn shape is linked genetically to other characteristics that were sought for in breeding.

Although the range of the bezoar goat is large, it seems to have been domesticated earliest in close proximity to the Old Mesopotamian and Levant cultural centers. Half a millennium elapses between the domestic goats found in Jericho and Jarmo and those found in Belt Cave levels near the Caspian. Elsewhere in the Near East domestic goats are generally found in the earliest phases of the domestic animal record.

[4] Frederick E. Zeuner, "The Goats of Early Jericho," *Palestine Exploration Quarterly* (1955), pp. 70–86. Cross-sections of horn cores of typical domestic and wild goats are also shown in Charles A. Reed's "Animal Domestication in the Prehistoric Near East," *Science*, 130, No. 3389 (December 11, 1959), 1629–39, see Fig. 3 on p. 1634.

[5] Reed, "Archaeological Evidence," pp. 130–31.

[6] *Ibid.*, p. 130.

[7] Charles A. Reed, "Osteological Evidences for Prehistoric Domestication in Southwestern Asia, "*Zeitschrift für Tierzüchtüng und Züchtungsbiologie*, 76, No. 1 (1961), 35.

The situation differs closer to India. In Anau domestic goats appear in the upper layers only, with sheep appearing earlier than goats. In the Indus Valley civilization of Harappa and Mohenjo Daro ca. 3000 B.C. there is no trace of goats although domestic sheep occur in the deposits. Present evidence does not allow the assumption of an African domestication of goats, but rather points to their introduction as already domesticated animals.[8]

SHEEP. The origin of domestic sheep, *Ovis aries aries*, is a more difficult problem than that of goats. There is an enormous taxonomic problem regarding the wild sheep of Eurasia, for there are many identifications of varieties and little agreement on the relationship between them.[9] In any event the genus *Ovis*, to which our sheep belong, includes species that today are mainly mountain, highland, and hill animals, as well as species that ranged widely over plains. The area occupied by the different varieties of wild sheep extended in mid-Pleistocene times from central Asia through the Near East and into Europe as far as England and France, as well as south through the islands of the Tyrrhenian Sea to North Africa. This range was sharply curtailed, probably by events connected with the last glaciation. In Europe wild sheep only survive in Sardinia and Corsica where the European mouflon, *Ovis aries musimon*, may still be found. While the European wild sheep have been isolated, those of the Near East are still in contact with the central Asia range. The most stately varieties survive in central Asia; outside of this area sheep equally large are only known in the fossil record. The whittling away of the area of wild sheep, chiefly by man, has led to the concentration of these animals in hilly or mountainous terrain. As a result wild sheep are often considered to be "mountain animals." Yet as Reed points out, mountains are essentially a retreat for wild sheep rather than their natural habitat.[10]

Since the earliest evidence for domestic sheep comes from western Asia,[11] it seems probable that the ancestral sheep of the west Asian domesticate is an Asian Mouflon, *Ovis aries orientalis*. Perhaps the particular animal was a geographic race of the mouflon *Ovis aries laristanica of Luristan*. It is a relatively short-horned, small animal of slender skull with a long mane. The attribution of ancestry to this animal rests in part on the similarity it bears to some domestic African sheep kept by primitive tribes.[12] A domestic animal strongly resembling it is depicted on cylinder seals from Warka, and similar representations persisted in Mesopotamia through Sumerian times. The Mesopotamian do-

[8] Reed, "Archaeological Evidence," pp. 133–34.
[9] *Ibid.*, p. 134.
[10] *Ibid.*, p. 135.
[11] Charles A. Reed, probably the leading student of animal domestication, believes that the known time-span for domestic sheep is of the order of about 11,000 years. The work of Dexter Perkins, to which Dr. Reed refers, in "Osteological Evidences," 34–35, may thus point to an earlier domestication of sheep than of goats. See our discussion of motives of domestication in Chapter 6.
[12] Caesar R. Boettger, *Die Haustiere Afrikas* (Jena: Veb. Gustav Fischer Verlag, 1958), p. 74.

mestic variety is spiral-horned and identical animals appear in the Gerzean site of Tukh in Egypt where they remain the dominant type of sheep throughout the Old Kingdom. There can be little question that this Egyptian sheep was of Asiatic origin.

But whatever the original sheep domesticated, it seems clear that various subspecies of wild mouflon mixed in the herds before the emergence of specific domestic breeds. Urial and argali sheep, whose range extends from the Punjab and Afghanistan westward over parts of Iran and into the Soviet Turkmen Republic as far as the Caspian, probably contributed to the early domestic stocks. The physical appearance of many present-day large-horned sheep of northwest India has also fostered the belief that they may be descended from a different wild ancestor, *Ovis aries cycloceros*, found today in the Punjab, Sind, and Beluchistan. In view of the significance attached to horn shape as index of domestication, it is interesting that this wild variety has spiral horns terminating in lateral points, often with an incipient second spiral.

If the above is a reasonably accurate description of what happened, then the earliest sheep domestication would have occurred on the eastern and central Asian periphery of the Near East. Other more western subvarieties may have subsequently entered into the prodomestic stock of Near Eastern sheep keepers, although Boettger argues that this is unlikely because sheep in a clearly domestic stage appear in early deposits of Near Eastern farming sites, and the incorporation of other wild types would have seemed neither necessary nor desirable.[13] The wild sheep of Corsica and Sardinia presumably did not contribute to the development of domestic breeds. Nor do the wild Barbary "sheep" of Africa seem to have played any part in the development of domestic varieties. Shape of horns, structure of skull, mane, and certain characteristic glands all differ from those of both domestic and wild sheep. Attempts at crossing Barbary sheep with domestic sheep have proved unsuccessful.

With the emergence of domestic sheep whose wild ancestors were both steppe and mountain animals, the penetration and settlement of very marginal grasslands became possible. It is likely that extensive pastoral nomadism developed only after the successful domestication of sheep by earlier farmers. Psychologically the sheep is perhaps the most domesticated of herd animals, depending completely on man for survival. The remarkable intelligence of wild sheep is completely lost in their domestic descendants.

CATTLE. Present-day types of domestic cattle, zoologists believe, are all derived from one ancestral strain, *Bos primigenius*, or the wild urus, an animal which survived in Europe until the late Middle Ages (the last remaining specimen died in 1627). *Bos namadicus*, whose relics are found in Asia and *Bos opisthonomous* found in North Africa are assumed today to be *Bos primigenius*. The urus formerly ranged over a vast area of Eurasia, from the Pacific to the Atlantic. The range also included northern Africa. The great area occupied by the urus accounts

[13] *Ibid.*, pp. 75–76.

for minor differences in the animal in different parts of its realm, and the names denote no more than geographic races of the animal.

Early cattle remains reveal considerable differences in size. Fossils indicate that the urus was a large, long-horned and powerful animal. Remains have been found within the period of early domestication revealing that the animal stood about seven feet high at the withers. But alongside remains of these enormous animals, fossil remains of considerably smaller cattle have been found throughout Europe, North Africa, and Western Asia. While the diversity in size has been interpreted in more than one way, recent studies have confirmed the view that the small animal was probably the female of the urus, the size difference largely disappearing in domestication.[14]

Apart from the osteological record, the earliest indication of the development of a domestic type distinct from the urus is in art. Representations suggest that the earliest strains of domestic cattle strongly resemble the urus. In many cases, of course, it is difficult to determine whether the animal portrayed was wild or domestic. On the famous standard of Ur, a bull is shown: that it was domestic may be inferred from the ring through his nose. Other representations, such as the victory tablet of King Narmer of Hierakonpolis (Kom el-Ahmar, Egypt), undoubtedly depicted wild bulls. However, much of the representational evidence, especially hunting scenes, is ambiguous. We may infer that the scenes depicting the hunt of cattle by Asshur-nasir-apal II (884–860 B.C.) show the hunting of wild cattle from the existing lists of game killed and captured.[15] On a single hunt this king killed fifty urus bulls and captured eight live ones. From other Assyrian texts we learn that young cattle captured in the hunt were bred in captivity. In the existing lists, different symbols are used for wild and for domestic animals, but the representations alone do not show conclusively that the animals were wild and not semidomestic cattle kept on the open range. Even the animals shown in hunts, or trapped in nets and through the use of decoy cows, may well be from herded range cattle. Boettger points out that the capture of bulls shown on two gold cups found at Vaphio near Sparta and dated between 1500 and 1250 B.C. is probably the capture not of wild bulls, as was previously supposed, but of semidomestic bulls.[16] Again, the long-horned massive cattle depicted in the bull games of Cretan frescoes are probably domestic, for they are pied.[17] Indeed, on a late Minoan picture, a cow of the same massive configuration is shown being milked in the old and dangerous Mesopotamian fashion—through her hind legs (a method that continues to be employed in Africa).

Representational art may be misleading for another reason. Styles persisted when they ceased to convey an accurate picture of the cattle

[14] Erich Isaac, "On the Domestication of Cattle," *Science* 137, No. 3525 (20 July 1962), 197. See further references cited in this article.
[15] Hanns von Lengerken, *Der Ur und seine Beziehungen zum Menschen* (Leipzig, Akademische Verlagsgesellschaft Geest & Portig K.-G., 1953), pp. 43–44.
[16] Boettger, *Haustiere*, p. 49. Cf. Zeuner, *Domesticated Animals*, p. 232.
[17] Lengerken, *Der Ur*, p. 42.

of the period. The maintenance of conventions characterizes religious art in particular. In Austrian churches, until recently, peasants offered statuettes of long-horned cattle, although such cattle had been unknown in Austria for many centuries. Wolf Herre's comparative study of skeletal remains and of pictures of domestic animals contemporary with the remains from medieval Hamburg revealed that very different conclusions would be drawn from the study of either alone.[18] That Egyptian representational art was characterized by the same maintenance of artistic conventions has been pointed out by J. Boessneck.[19] It is of course possible to identify domestic cattle in representations that stress specifically domestic characteristics—for example, pied coats, extremely large udders, or short-horned or polled cattle—such as are found on the mural relief of King Ti at Saqara, Egypt, of the twenty-fifth century B.C.

From the wild urus two races of domestic cattle emerged early. In one, the heavy horns of the urus caused the development of wide and flat parietal bones, so that the top of the skull appears almost flat when the animal is seen head on. Domestic cattle that retained a urus conformation of skull and body are called "primigenius" cattle, *Bos taurus primigenius,* and are the earliest domestic cattle. When short-horned domestic cattle developed, the frontal and parietal bones, released from the excessive weight of horns, became domed; this is most evident in polled animals. This type, because of its characteristic long and narrow face and upward convex parietals, is called *Bos taurus longifrons.*

Longifrons cattle, differing markedly from primigenius cattle, like the latter first appeared in Mesopotamia. While it is difficult to distinguish between urus (*B. primigenius*) and domestic primigenius cattle (*B. taurus primigenius*) in the early Mesopotamian representations, in the case of longifrons it is clear that a domestic type is represented. Moreover, longifrons cattle are generally depicted in association with agricultural performances or symbols. Probably the first representation of longifrons is on a bowl of the Jemdet Nasr period; longifrons are depicted more often thereafter, although never as frequently as primigenius. Caesar R. Boettger has proposed that the distinction between longifrons and primigenius was one between an economically exploited breed and a strain maintained primarily for ritual purposes. The distribution of longifrons cattle outside the Near East is taken by Boettger to indicate that they were spread intentionally and did not originate independently in a number of places. In spite of the fact that longifrons cattle appear much later than primigenius in Mesopotamia, their docility and over-all usefulness, according to Boettger, account for their having reached both the Atlantic and the Pacific ends of the Old World before primigenius cattle.[20]

[18] Isaac, "Domestication of Cattle," p. 199.
[19] "Die Haustiere in Altägypten," *Veröffentlichungen der zoologischen Staatssammlung München,* 3 (1953), 1–50.
[20] Isaac, "Domestication of Cattle," pp. 199–200. The development of zebu cattle cannot be treated in detail in this study. The reader is referred to various studies by H. Epstein, e.g. "Descent and Origin of Afrikander Cattle," *Journal of Heredity,* XXIV, No. 12 (December 1933), 449–62. "Phylogenic Significance of

PIGS. The ancestors of the domestic pig, *Sus scrofa domesticus*, belong to a species *Sus scrofa* which ranges from New Guinea and the Pacific margins of Eurasia westward through Europe and Northern Africa to the Atlantic. Contrary to widely held notions, the wild pig is not primarily a forest animal. The pig is found wherever in his range his requirements for water and concealment are satisfied, whether in forest, steppe, or even semidesert, or in deserts near spring, rivers, or marsh.[21] There are many subspecies confined to specific areas, but these are closely related and their continuity with the subspecies of contiguous areas is evident. It seems that ecological conditions and local breeding range are responsible for the differences. Once wild pigs were domesticated, local subspecies probably were brought into the stock to contribute to the formation of domestic breeds. Such bringing in of wild stock was done intentionally until relatively recently in Europe, and is still done today in southeast Asia. Inadvertently such crossing would also have occurred when wild boars intruded into the domestic herds. (One reason for introducing local wild boars to domestic pigs could be to balance disturbed reproductive functions of the domestic animal. When pigs are raised for meat and lard, mutants with endocrinal disturbances encouraging fat deposition but interfering with reproduction often become important in the stock.) [22]

In view of the wide range of the wild ancestor of the pig and, until recently, the resemblance of local domestic to local wild pigs, it is difficult to locate the hearth of domestication. I believe a western Asian origin of pig domestication is the best hypothesis in terms of the available evidence. It seems definite that an original domestication did not occur in Europe, for domestic pigs appear there rather suddenly and in conjunction with other domestic animals. The Wei and Hwang Ho region in east Asia, the hearth of Chinese civilization, is unlikely to have been either an original hearth or an independent center of domestication. The domesticated pig is found in numerous early sites, but its skeletal features are not similar to those of the wild species of China.[22a] And while some students have considered Southeast Asia the original hearth of pig domestication—partly because of the many varied methods of pig keeping, including the keeping of semidomestic pigs, and the close contact often found between wild and domestic animals [23]—there is no archaeological evidence for this belief, despite fairly intensive investigation of sites in Southeast Asia. Domestic pigs only appear around 2000 B.C. in the Somrong-Sen culture, which extended from the lower course of the Ganges southeastward over the entire Indo-Chinese

Spina Bifida in Zebu Cattle," *Indian Journal of Veterinary Science and Animal Husbandry*, XXV, Part 4 (December 1955), 313–16. "The Origin of Africander Cattle, with Comments on the Classification and Evolution of Zebu Cattle in General," *Zeitschrift für Tierzüchtung und Züchtungsbiologie*, 66, No. 2 (1956), 381–96.

[21] Reed, "Archaeological Evidence," p. 139.

[22] Boettger, *Haustiere*, p. 21.

[22a] William Watson, "Early Animal Domestication in China." In Ucko and Dimbleby, *Domestication and Exploitation of Plants and Animals*, pp. 393–95.

[23] Cf. Sauer, *Agricultural Origins*, pp. 31, 33, 37. Boettger, *Haustiere*, p. 23.

peninsula and as far as the Yunan plateau and Taiwan. The culture first appears gradually in the west, where it has links with other western cultures, but enters full blown in the east, all of which suggests a western origin for the culture and an introduction of pig keeping, along with the culture to which it was linked. Boettger suggests that the Somrong-Sen culture may have introduced pig keeping to Southeast Asia as well as to China.[24]

The archaeological evidence from Anatolia and Kurdistan is in favor of the Near East as the area of domestication of pigs, for the process of domestication seems to have been underway there by the middle of the seventh millennium B.C.[25] In later periods pigs became locally very important in Mesopotamia.[26] India, often treated as a possible center of domestication for pigs in the past, is no longer so considered, for there is a gap of nearly four thousand years before the first domestic pigs appear there. Wherever the hearth of the first Near Eastern domestication, the practice of pig keeping must have spread throughout western Asia and beyond, and local wild stock was drawn on as much as introduced animals.[27]

The attitude of early historic peoples of the Near East to pigs can only be understood if the animal is seen as present in the area for a long time.[28] The very fact that religious prohibitions against the pig were so important indicates its long history as a sacral animal regarding which, as new religious conceptions became prevalent, new attitudes developed.

OTHER BOVINES. Other important domestications occurred on the periphery of the Near Eastern hearth and, as in the case of plant domestication, it is possible to apply to them the concepts of secondary and substitute domestications.

At the eastern and southeastern fringes of the Near Eastern hearth a number of bovines were domesticated. These include the mithan or gayal, *Bos (Bibos) frontalis*, probably derived from the Malayan gaur, *Bos (Bibos) gaurus;* the banteng of the Greater Sundas, derived from the wild *Bos (Bibos) banteng;* the Tibetan yak, *Bos poephagus grunniens;* and the water buffalo, *Bubalus bubalis*, the domesticated variety of *Bubalus arnee*. Mithan, banteng, and yak have not spread as domesticates beyond the area of their ancestral wild stock. They were probably south and southeast Asian substitute domesticates for cattle in areas where

[24] Boettger, *Haustiere*, pp. 23–25.

[25] Reed, "Patterns of Animal Domestication," p. 371.

[26] Max Hilzheimer, *Animal Remains from Tell Asmar*, Adolph A. Brux, trans., Oriental Institute of the University of Chicago, Studies in Ancient Oriental Civilization, No. 20 (Chicago: University of Chicago Press, 1941).

[27] For a discussion of the argument concerning derivation of domestic pigs see Karl Hescheler and Emil Kuhn, "Die Fauna der neolithischen und metallzeitlichen Siedelungen," in Tschumi, *Urgeschichte*, I, 293–94. See also Reed, "Archaeological Evidences," p. 139.

[28] On religious food prescriptions see the comprehensive study by Frederick J. Simoons, *Eat Not This Flesh; Food Avoidances in the Old World* (Madison, Milwaukee and London: The University of Wisconsin Press, 1967). On pigs and pork see pp. 13–43.

cattle did not thrive, and perhaps were the product of a general movement by peasants to domesticate animals similar to cattle. These domestication attempts, although successful, did not interfere with the continued introduction of cattle into these areas, nor indeed did they impede the advance of domestic cattle toward the tropics except in areas of decisive environmental unsuitability.

Some scholars believe the bovids that have been termed here substitute domesticates were domesticated first and that cattle were substitute domesticates.[29] The chief argument against this is that the direction of agricultural dispersal is overwhelmingly from within the Near Eastern hearth outward. Moreover, the other domesticated bovines generally remained confined to areas in which the wild species thrived. Their domestication thus bears the stamp of substitute domestication. Only the domesticated water buffalo, probably the oldest of these secondary domesticates, spread widely from its hearth of domestication in the Indian or southern Mesopotamian sector of the Near Eastern periphery.[30] Formerly, the range of the wild water buffalo was large, almost coinciding with that of the urus. But the buffalo continued to be hunted, in fact hunted out, over much of the western part of its range in North Africa and western Eurasia (including the Near East) long after cattle are definitely known to have been domesticated.

Domestic buffalo were apparently re-introduced into the Near East from India after the disappearance of local stock. The Greeks met the animal when they penetrated Asia, but buffalo were only introduced into Egypt in the early Islamic period. The great importance of the animal in Egypt dates back only to the latter part of the last century, when cattle were almost entirely annihilated by the rinderpest epidemic of the 1890's. It is mainly in southwest and southeast Asia, including China and Japan, that the water buffalo has become an important domestic animal. The animal is particularly well suited to moist areas such as the rainy tropics and is of course widely used there for traction, plowing, and carrying loads. The sacred use of milk and cattle as well as avoidance of meat, which is often believed to have accompanied the expansion of cattle-keeping peoples into the southeast Asian tropics, was extended to water buffalo.[31]

THE ASS.   The ass, *Equus asinus*, in Boettger's view is the earliest economically important animal derived from the African fauna.[32] On a

---

[29] B. Klatt suggested that the water or arni buffalo was the first bovine to be domesticated in Mesopotamia. *Haustier,* p. 34.

[30] Max Hilzheimer is believed to have Mesopotamian representational evidence of domestic water buffalo from ca. 2000 B.C., *Die Wildrinder im alten Mesopotamien,* Mitteilungen der Altorientalischen Gesellschaft, 2, No. 2 (Leipzig: Verlag von Eduard Pfeiffer, 1926), p. 14. The animal was almost certainly considered domestic in the biblical period (Samuel II, 6:13; Isaiah 1:11).

[31] Paul Wheatley, *A Note on the Extension of Milking Practices into Southeast Asia during the First Millennium A.D.,* Southeast Asia Reprint Series, No. 219 (Berkeley, Calif.: University of California Center for Southeast Asia Studies, Institute of International Studies, 1965).

[32] *Haustiere,* p. 106.

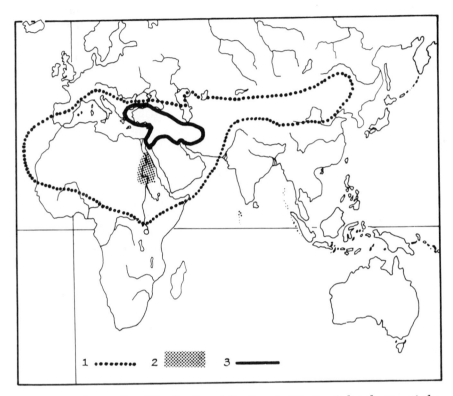

Figure 8. *Premodern Distribution of the Ass. 1, Limits of distribution of the ass; 2, Presumed area of domestication of the ass; 3, Oldest hearth of plant and animal domestication.*

slate palette dated to the Gerzean (end of the fourth millennium B.C.) asses are shown together with domestic sheep and cattle and it can therefore be assumed that the asses were also domestic.[33] The ass derives from the African wild ass which had a more northerly distribution than any other African equid. The Nubian variety seems to have been domesticated first, although other wild races came to be incorporated into the domestic stock. The range of the wild ass extended in the past from Somaliland through Nubia and the Saharan uplands to the Atlas, and its domestication may perhaps have resulted from contact with cultures entering from Asia that already had domestic herd animals.

It is impossible to reconstruct the process by which the ass was domesticated. It was hunted in North Africa for food, although in recent times ass meat has lost favor. One may conjecture that pet animals were sometimes allowed to survive and were substituted for cattle as beasts of

[33] Reed, "Archaeological Evidence," p. 144.

burden. There may have been a pattern of development in which the tasks of the major herd animals were seen to be better fulfilled by another animal, at least in certain areas. Thus sheep were early used as beasts of burden (a use that persisted until the twentieth century in western Tibet).[34] Cattle, vastly superior to sheep for this purpose (at least where they can adequately be watered), were widely substituted for sheep as carrying animals. The ass was then substituted for cattle and/or sheep in dry areas, and the camel eventually took the long distance carrying function over from cattle in the drylands of the Old World.

Domestication of asses may also have been encouraged by the practice of keeping herds of the animal for the royal hunt, important through the latter half of the second millennium B.C. in Egypt, especially in Egyptian controlled parts of Nubia. This use of asses may explain the very large herds that were kept in the fourth dynasty (2840–2680 B.C.). This was also the period of great pyramid building, and asses may have been used as beasts of burden in construction works. While they may have been used to carry, it is doubtful that asses were used for pulling to any large extent, for the inappropriate harnesses derived from cattle harnesses that were used on asses reduced their effectiveness in traction by at least two-thirds of modern expectations. Lack of proper shoeing also reduced the effectiveness of work animals. Finally, and perhaps most importantly, the greater power of the ass in relation to man is partly cancelled out by the greater cost of maintaining him by the standards of an ancient economy.[35] In short, there was no economic incentive to prefer draught animals over slaves or the corvée and it may be that the herds of asses were kept in Egypt for prestige, ritual and military purposes rather than strictly for economic reasons.

The domestication of the ass led to a considerable expansion of nomadism. The animal was suitable for carrying women, infants, and old people, as well as camping gear and other moveables. The frescoes of Beni Hasan (1892 B.C.) depict Semitic nomads accompanied by asses which carried children and goods.[36] Indeed, in the patriarchal age Near Eastern trade depended on the ass for carriage and the Biblical patriarchs rode on asses (Genesis 22:3).

Nonetheless, there are significant limitations to the use of asses. These permitted their use on the march, but not in battle. Asses are more independent than horses and, as befits the hilly terrain of their wild range, more deliberate. It is therefore much easier to train horses to submit to man even in conditions of battle, for horses are psychologically disposed to submerge their individual will. It would be dangerous to ride an ass in battle, for the animal might become recalcitrant in conditions where it might be fatal for a rider to fight his mount as well as the enemy. As Boettger points out, historically we do not find an ass-mounted cavalry

[34] Burchard Brentjes, *Wildtier und Haustier im alten Orient* (Berlin: Akademie-Verlag, 1962), p. 21.

[35] R. J. Forbes, *Studies in Ancient Technology* (Leiden: E. J. Brill, 1965), II, 81–88.

[36] Cf. W. F. Albright, *The Archaeology of Palestine*, p. 209.

at any time.[37] The ass was in fact a symbol of peace, and this was the reason for the biblical prophecy concerning the Messianic king riding upon an ass (Zechariah 9:9-10). In this famous passage the ass is specifically contrasted with the horse, the symbol of war.

Toward the end of the fourth millennium B.C. asses were introduced via Egypt to the entire Near East. They appear subsequently in Sumer and their rapid spread is presumably due to nomads. The ass was introduced to Asia Minor and Greece, and became important in Italy in the latter half of the first millennium B.C. By the Greco-Roman period asses had become important work animals in all Mediterranean countries.

THE ONAGER.   The onager, *Equus hemionus onager*, is a relative of the ass, but the distribution of the wild animal is in central, western and southwest Asia. The wild onager ranges from northern Mesopotamia through the Iranian plateaus, Turkistan, Uzbekistan, and Tadzhikistan into northwestern India.

In ancient Mesopotamia the onager was clearly a substitute domesticate. It was harnessed both as a pack animal and for traction, in the manner of cattle. The methods of control earlier used for cattle, such as nose-ring, ropes, and headstraps, were supplemented in the case of the onager with a curb or noseband which restrained it from biting. Onagers were used widely in Mesopotamia and the Iranian–Mesopotamian border zones for carrying, for traction, and for chariot teams. They continued to be hunted for food as well as for addition to the stock of domestic animals, for it was apparently difficult to establish truly domestic strains of onagers.[38] In the Syrian–Palestinian area onagers were unable to compete seriously with cattle in traction or as pack animals.[39] Indeed, by the time the onager was widely used in Mesopotamia, domestic asses introduced from Africa had taken over the pack carrying function and thus had preempted a major purpose which might have provided a rationale for a continuation of onager husbandry. Other domesticates (camel, dromedary, horse) also served to deprive onagers of their economic role.

THE BACTRIAN CAMEL.   The range of the Bactrian camel, like that of the onager, was the Asian periphery of the Near Eastern agricultural centers. This animal's range overlaps that of the wild onager, except that it extends further into the arid heart of Asia.

The domestication of the camel allowed an extension of trade across deserts and a further development of nomadism, hitherto limited by the physiological requirements of the sheep and ass. However, the period,

---

[37] *Haustiere*, p. 109.

[38] F. S. Bodenheimer, *Animal Life in Biblical Lands* (Jerusalem: Bialik Institute, 1956), I, 77, 138–39, 158–59, and *passim*. Ironically, had the domestication of onager been successful the animal might have been protected from extinction. An extremely harsh winter about fifty years ago led to their complete extinction in the Syrian desert. Jehuda Feliks, *The Animal World of the Bible* (Tel-Aviv: Sinai, 1962), p. 29.

[39] The Talmud mentions that onager were used to turn millstones (*Abodah Zara* 16b). It also reports that they were fed to the beasts in the imperial arena (*Menahot* 103b).

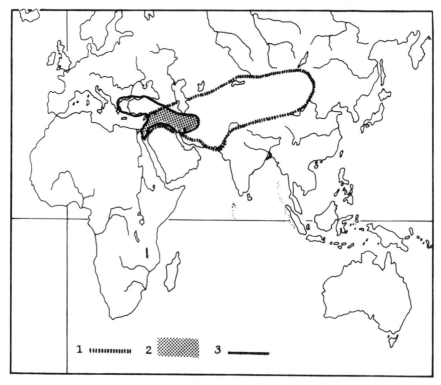

Figure 9. *Natural Distribution of the Onager. 1, Limits of the range; 2, Presumed area of domestication of the onager; 3, Oldest hearth of plant and animal domestication.*

purpose, and method of the Bactrian camel's domestication remain controversial. Fossil bone studies show that the camelid family was once widely distributed from the New World through central Eurasia and throughout North Africa. Toward the end of the Pleistocene camels withdrew from North Africa, just as they had disappeared earlier from the western hemisphere, where the family had been well represented in older geological periods. Possibly hunting and/or the late pluvial of the region contributed to the disappearance of camels in extensive areas of North Africa.[40]

In the past many writers did not distinguish sufficiently between the dromedary or one-humped camel, *Camelus dromedarius*, and the Bactrian two-humped camel, *Camelus bactrianus*. The dromedary is more tolerant of heat and its domestic varieties dominate in the North African,

---

[40] Marvin W. Mikesell, "Notes on the Dispersal of the Dromedary," *Bulletin de l'Institut Français d'Afrique Noire*, XVIII, series A, No. 3 (July 1956), 896–98.

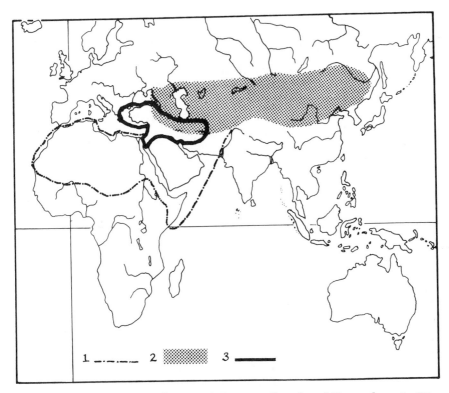

Figure 10. *Natural Distribution of Bactrian Camel and Dromedary. 1, Distribution of the dromedary; 2, Distribution of the Bactrian camel; 3, Oldest hearth of plant and animal domestication.*

Levantine, and Arabian parts of the Islamic world. The earlier theory that the dromedary is only a domestic variety of the Bactrian camel is probably wrong. The animals seem to have different wild ancestors, and hybrids, although fertile, bear characteristics of interspecific hybridization. The hump of offspring of Bactrian males and dromedary mares is usually separated into a dwarfed anterior and a more pronounced posterior part. Such animals normally surpass both parents in endurance and size.[41]

Phylogenetically the Bactrian camel is the more primitive type, retaining two-humpedness as an adult feature, while in the dromedary the double hump, occurring in the embryo, is fused into a single hump before birth. Thus the dromedary is a more recent member of the Camelidae.[42]

[41] Boettger, *Haustiere*, pp. 214–15.
[42] On the development of the Bactrian camel see Reinhard Walz, "Neue Untersuchungen zum Domestikations-problem der altweltlichen Cameliden. Beiträge zur Geschichte des zweihöckrigen Kamels," *Zeitschrift der Deutschen Morgenländischen Gesellschaft*, 104, No. 1, n. ser. 29, (1954), 45–87.

Single-humpedness in camels is not likely to be the result of domestication, which encourages the persistence of infantile features, in this case two humps. On the other hand might the Bactrian camel be a domestic variant of wild dromedary with the infantile characteristic of double-humpedness retained as adult features? Were that the case, the surviving wild camels of central Asia, which are double humped, would be feral. They have been treated as such by some: formerly, for example, wild Dzungarian camels were classified as feral. Boettger argues that, were this view correct, feral animals would in the course of time have reverted to a one humped form, for domestic characteristics are lost in the wild state over time.[43] It now seems reasonably certain that the dromedary evolved from an extinct southern wild variety or varieties, and the Bactrian camel from northern wild stock.[44]

Bones of Bactrian camels have been found from India through central Asia and Siberia to the Volga river, and wild Bactrian camels, more like fossil varieties than modern camels, still occur in the Tarim Basin. Unfortunately wild dromedaries have not survived, although Agatharchides of Cnidus (late second century B.C.) claimed that in his day there were still wild dromedaries in Arabia. There seems to have been a fairly wide transitional zone between the range of wild dromedary and wild Bactrian camels. Diodorus (ca. 100–20 B.C.) mentions domesticated Bactrian camels in Arabia and it is certainly possible that wild Bactrian camels were also found there. Similarly, judging by ancient texts, domestic dromedaries were known north of their modern distribution. Whether the Avesta refers to camel or dromedary cannot be determined. Nonetheless it is fairly certain that wild Bactrian camels preceded dromedaries in northern Iran, a region which in the opinion of classical authors is the homeland of the Bactrian camel.[45] Bone remains of the dromedary have been found in North Africa and in southern Europe, generally west and south of the area of Bactrian camel bone finds.

The Bactrian camel was probably domesticated in the third millennium B.C. in central Asia.[46] Forbes' contention is that the animal was domesticated by people who had earlier domesticated cattle and who were subsequently to domesticate horses.[47] (The domestication of the related dromedary is discussed after the domestication of the horse because I believe it was the development of riding which made the dromedary a desirable substitute domesticate for the Bactrian camel.)

THE HORSE.   The horse is the first animal to which riding techniques were applied. The earlier domesticates had carried both goods and people, but riding as a specific technique was an invention of the domesticators of the horse.[48]

---

[43] *Haustiere*, p. 215.

[44] C. R. Boettger, "Zur Frage des Erlöschens gewisser Tierarten nach deren Domestikation," *Sitzungsberichte der Gesellschaft Naturforschender Freunde zu Berlin 1937*, No. 8/10 (1938), 320–24.

[45] Forbes, *Ancient Technology*, II, 195.

[46] *Haustiere*, p. 216.

[47] *Ancient Technology*, II, 193.

[48] This section draws heavily on Franz Hancar's comprehensive study *Das Pferd*, *op. cit.*

It was formerly believed that the domestic horse derives from a wild central Asiatic horse called Przewalski's horse, *Equus caballus prezewalskii* which lives, or used to live until recently, in western Mongolia and eastern Turkestan. This horse is similar to small Mongol horses, but it has been shown that morphological and other characteristics of Przewalski's horse (e.g., short erect mane, no forelock) differ sharply from those of Mongol horses. The latter are clearly derived from the tarpan, the so-called "European wild horse," *Equus caballus ferus*, which may have survived until the twentieth century in eastern Russia, but is now extinct. The range of the tarpan extended from western central Asia to central Europe, but it was first domesticated in the eastern part of its range, where its distribution touched and in part overlapped that of Przewalski's horse. Presumably first attempts at domestication were made in the third millennium B.C. The first known osteological finds are from the archaeological Afanasevo culture of the Kuznetsk region and the Andronovo culture (third-second millennium B.C.) of the steppes north of the Aral Sea in central Asia. It is only in the second half of the second millennium B.C. that horse bones are found in substantial numbers west of the Ural River.[49]

The horse was probably a substitute domesticate performing labor services similar to those rendered by cattle, Bactrian camels, or onagers.[50] Early artistic representations of horses show people being carried as loads, squatting in various positions on the animal and not yet riding. Many local types of horses developed, due partly to the fact that numerous geographic races of tarpans were drawn upon for the formation of domestic herds, and wild blood was undoubtedly repeatedly reintroduced by wild stallions breaking into semidomestic herds. Eventually two dominant geographic races developed: a long-haired, heavy northern, and a short-haired, light and fast southern stock. The northerly tarpans are ancestral to the Mongol horses and those subsequently introduced into China. The southern varieties have become the most important domestic horses and their development culminated in the "hot-blooded" horses represented in their purest form today by the Arabian horse. The pure-bred Arabian horses, insignificant in number today, are historically important, for their blood flows in the veins of almost all thoroughbreds and has been introduced into most breeds.[51]

The impact of the southern tarpan races results from the fact that horses derived from them were widely dispersed in the wake of the marauding and conquering mounted peoples who moved out of central Asia and broke into the Near East. The Near East had few wild horses—

[49] The derivation of domesticated horses from the tarpan is challenged by Günter Nobis who believes that Przewalski's horse may after all represent the ancestral strain. "Beiträge zur Abstammung und Domestikation des Hauspferdes," *Zeitschrift für Tierzüchtung und Züchtungsbiologie* 64, No. 3 (1955), 201–46.

[50] James F. Downs, "The Origin and Spread of Riding in the Near East and Central Asia," *American Anthropologist* 63, No. 6 (December, 1961), 1193–1203. It is Down's view that horseback riding could only develop after the size of the horse was increased through selective breeding.

[51] Cf. George Gaylord Simpson, *Horses*, The Natural History Library published in cooperation with the American Museum of Natural History (Garden City, New York: Anchor Books, Doubleday & Company, Inc., 1961), 52.

indeed over large areas, none at all. Hence no variations were introduced by hybridization and conformation was maintained. The horse arrived in Mesopotamia shortly before 2000 B.C. It was probably introduced via Turkestan and Iran, but its importance could not have been great, for in the oldest law codes, including Hammurabi's Code (1728–1686 B.C.), there are numerous laws pertaining to other domestic animals, but no mention whatever of horses. There were no horses in the Indus Valley cultures of Harappa and Mohenjo Daro or in the older deposits of Sumer. There is earlier evidence for their introduction into Anatolia both in texts and in remains of funeral offerings in archaeological deposits from the eighteenth to the fourteenth century B.C., but the animals were rare and kept only by the nobility. Only the joining of horse and wagon led to widespread acceptance of horses late in the second millennium B.C. The horse appeared in Egypt ca. 1700 B.C. with the Hyksos conquerors who used it for their chariots. The harnessing of horse to war chariot resulted in an extremely wide distribution of both. In the course of the second millennium, horse and chariot spread as far as the Niger in Africa, India and China in Asia, and Scandinavia in Europe. The horse as a major tool of war remained important until the twentieth century.

It is not immediately easy to see how the horse came to be accepted by the peoples of the ancient Near East. The horse was not better suited for long distance transport than camels or asses, and for short hauls bulls and oxen were vastly superior. It seems probable that the ceremonial of sacral Near Eastern kingship as well as sheer love of public display and ostentation made royalty consider the as-yet very rare horse a proper symbol of might.[52] In the Mari kingdom of northern Mesopotamia the horse was not yet important in warfare but important in royal ritual. Thus Zimri-Lim (1722–1690 B.C.) is advised not to use horses for fear of offending the Akkadians. Before that Shamsi-Adad (1744–1724 B.C.) wanted to borrow Yasmah-Adad's famous chariot team in order to impress a foreign delegation present at his *akitu*, a ritual accession and New Year celebration. The value placed on horses by royalty was very high. Ishi-Adad of Qatna offered horses that he valued at 600 silver shekels to Ihsme-Dagan (1723–1693 B.C.). J. Sasson points out that a horse was thus worth fifteen oxen and an equal if not larger number of asses.[53]

The European breeds were derived locally from southern and northern tarpans introduced by mounted hordes. These crossed with local geographic races of tarpans to produce the European "cold-bloods." The European horses at first gave rise to a population with no complete segregation of types, but eventually numerous ponies, especially the "Celtic pony," ancestor of today's Iceland pony, as well as large and heavy breeds

[52] In the Bible the horse is usually regarded as a mount for war and Deuteronomy (17:16) warns the king against the despotic potential of mounts: "Only he shall not multiply horses to himself. . . ." Nonetheless it became a ceremonial animal in Israel as elsewhere, especially in the paganizing tendencies of late Judean kingship. King Josiah (637–608 B.C.) eradicated horse-worship (Kings II, 23:11).

[53] Jack M. Sasson, "A Sketch of North Syrian Economic Relations in the Middle Bronze Age," *Journal of the Economic and Social History of the Orient*, IX, part III (1966), 175–77, especially note 1, p. 176.

emerged. The medieval Great Horse, which carried heavily armed knights, was developed from the heavier horse. There was a continental or Flemish and a British type. From the former the modern heavy draft horse, including the Belgian and Percheron, developed; and from the latter, the Shire, Clydesdale, and Suffolk. In any event it is clear that the European horses, whether diminutive or massive, are closely related. Percherons, for example, have been successfully crossed with the small Shetlands.[54]

The development of horseback riding made it possible to use other animals for riding as well. Onagers were the first animals after the horse to be used for riding. The archaeological record makes it clear that bits and bridles were not applied to the onager, or to any draft animal, for a long time. Bridle bits are very old in central Asia, where there have been numerous finds of bits made of wood, bone, antler, and later of metal. The invention of the bit is so striking that its later association with other animals cannot be explained through repeated parallel inventions, but only through transfer from the horse.

In the ancient Near East, though highly valued, the horse never became common, presumably because of more limited natural pasture in that area and an established agricultural system that put horses in competition with man for grain. Its only important use was in the war chariot which became a major weapon in the course of the second millennium B.C. That the horse was not fully exploited in war can be explained by the absence of a saddle. For cavalry purposes the cattle-blanket did not allow a firm seat. As a chariot animal, the horse slowly replaced the onager. In Persia however, the onager continued to be harnessed to chariots at least until the fifth century A.D.; and in India, too, the onager was used in this fashion until the third century A.D. The extensive hunting of the onager, as was noted earlier, contributed to its decline.

DROMEDARY. The dromedary became the major riding animal of the Near East. Originally it must have been a substitute domesticate for the Bactrian camel. Boettger argued that two humped animals were difficult to obtain in the early period of their introduction to the Near East, with the result that a native and more accessible animal was substituted for a coveted but rare one.

But the dromedary owes its great importance to its use as a riding animal. The inspiration for this probably came from the central Asian horse nomads. The orthostat of a dromedary rider from Tell Halaf of approximately 2000 B.C. shows the rider wearing scale armor which is unquestionably of central Asian nomadic origin.[55] The raising of domesticated dromedaries spread quickly in Syria, Palestine, and especially in southern and central Arabia, an area which had large wild herds.[56] It is from the Arabian and Syrian border lands that Mesopotamian rulers

---

[54] Simpson, *Horses,* p. 59.

[55] Boettger, *Haustiere,* p. 217.

[56] Reinhard Walz believes that the domestication of dromedaries began in central Arabia in the second half of the second millennium B.C. "Zum Problem des Zeitpunkts der Domestikation der altweltlichen Cameliden," *Zeitschrift der deutschen morgenländischen Gesellschaft,* 101, No. 1, n. ser. 26 (1951), 29–51.

obtained quantities of dromedaries in peace and large numbers as booty in war. However, they only became important there in the first millennium B.C.

In pre-Israelite Canaan there were either dromedaries or camels, the archaeological evidence not making clear which they were.[57] In the Israelite period the association of these animals with desert dwellers is frequently mentioned. At the time of Gideon (ca. 1100 B.C.) their role in Midianite raiding is evident, but in warfare the camel assumed importance only after the Christian era.

The domestic dromedary never reached Africa in ancient times. There is no evidence that the Libyans knew the dromedary; indeed rock drawings from northwest Africa, Libya, Egypt, and Nubia that were thought to represent domestic dromedaries have been shown to be of relatively recent age. This is true for all the rock art between Aswan and Wadi Halfa, as well as in the Western Desert. The dromedary played no part in Jugurtha's war (106–104 B.C.), nor did it play any part in the Carthaginian wars. The evidence is strong that the animal in its domesticated state was not introduced into Egypt until the Greco-Roman period (ca. 300 B.C.). Assyrian and Persian conquerors of Egypt did bring dromedaries with them, but this apparently had no impact upon Egyptian animal husbandry. Libyan dromedary riders do not appear in desert warfare until the second century A.D.[58]

THE DOG. In the discussion of domestic animals it has long been customary to begin with the dog as presumably the first domesticated animal. Actually there is not sufficient evidence to support what has become a "dogmatic" assumption.[59] The process of domestication of dogs is extraordinarily difficult to trace. It is not possible on the basis of osteological characters to distinguish tamed wolves from early domestic dogs, though fully domestic dogs can be distinguished from wild wolves. Many archaeological finds thought to reveal transitional or early domestic forms, for example at Jericho, have been discredited.[59] Reasonably certain identifications of domestic dogs are reported from England Anatolia, and Kurdistan from ca. 7000 B.C. and from Idaho more than one thousand years earlier.[60]

A majority of zoologists believe that all the varieties of the domestic dog *Canis lupus familiaris* are derived from the wolf, *Canis lupus*. Both have the same diploid chromosome number; they interbreed readily and produce fertile offspring. There are no clear conformational characteristics by which smaller wolves, e.g., those of southwest Asia (*Canis lupus pallipes*), can be distinguished from dogs. Furthermore, young

[57] Mikesell, "Dispersal of the Dromedary," pp. 896, 908.

[58] *Ibid.*, pp. 903–5.

[59] Juliet Clutton-Brock, "Carnivore Remains from the Excavation of the Jericho Tell," in Ucko and Dimbleby, *The Domestication and Exploitation of Plants and Animals,* pp. 337–45. "Mesolithic" sites in which "dogs" were found have turned out to be much more recent than the Neolithic farming villages in which none have been as yet determined. Cf. Wolf Herre, "Der heutige Stand der Domestikationsforschung," *Naturwissenschaftliche Rundschau*, 12, No. 3 (March, 1959), 88.

[60] Reed, "Pattern of Animal Domestication," p. 370.

wolves tame easily. Conversely, the behavior of dogs, whether as individuals or in groups, can be traced back to wolves. Differences between
modern dogs and wolves can be explained by the extraordinary breeding
plasticity of the dog, which permits rapid change in conformation and
aptitude over a few generations.

Old views identifying the jackal as ancestor of the dog have been
abandoned. The chromosome numbers do not match, and even Lorenz,
a recent advocate of the jackal ancestry theory, has abandoned the view.[61]
A third view is that the dog is derived from a wild dog ancestor whose
surviving representatives include the Australian dingo and the pariah
dog of Moslem North Africa and some parts of Europe and eastern and
southern Asia. As Reed says,

The "wild dog" theory in one variant or another has had much support. Some
of this would seem to be emotional, on the part of dog-lovers who abhor
wolves and whose attitude is fostered by the survival in our culture of medieval European folk tales of the wolf. However, the theory also has several
serious supporters. . . . As Bodenheimer remarks, what is needed is a
thorough study of the neglected pariah, still found throughout much of the
Eastern Hemisphere but almost unknown zoologically because of the general
assumption that it is but a half-feral domestic dog.[62]

There are others who believe that different dogs are derived from different ancestors with some, e.g., greyhounds and salukis, coming from the
pariah-dingo species and others from the wolf.

Original contact between wolf and man is often said to have come
about when wolves attached themselves to human camp sites. One may
indeed ask, as in the case of the pig, "Who originally domesticated
whom?" The wolf may have been ignored and simply tolerated as a camp
follower until, in the course of time, a semidomestic relationship developed. In this view it seems reasonable to assume that there were repeated
domestications of young wolves, perhaps raised as pets for no particular
purpose except to be eaten occasionally. The utility of these dogs became
apparent as an alarm against enemies or as an aid in flushing out game.

In its general outlines such an account of domestication of the dog
may well be correct. It rests, of course, on certain assumptions, such as
the experience with wild dogs of the pariah or dingo variety who roam
in packs around a village or camp. It also rests heavily on ethological
experience with wolves. Nonetheless the likelihood must be considered
that the dog was first domesticated in one place, in a geographically
circumscribed part of its range, and that domestication of dogs spread
from there. The southwest Asian limit of the distribution of wolves has
a number of small varieties, e.g., *Canis lupus pallipes*, and dogs very
similar to these wolves appear quite suddenly in European sites, which
may indicate that they were introduced and not derived from local wolf
stock. Perhaps only after introduction of domestic dogs did crossbreeding

[61] Reed, "Addendum" on p. 126 of "Archaeological Evidences."
[62] Reed, "Archaeological Evidences," p. 127.

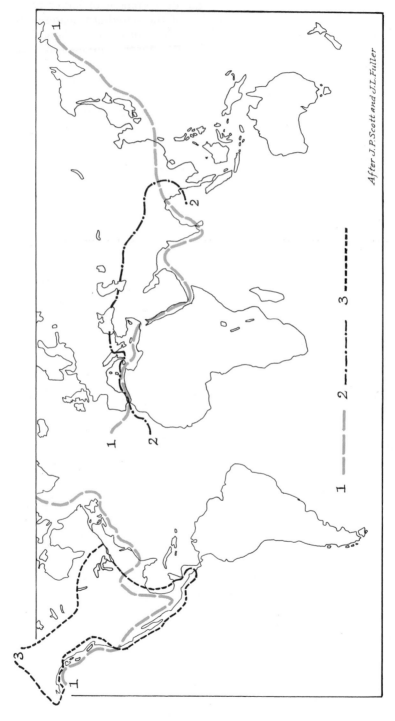

Figure 11. *Geographical Distribution of Members of the Genus Canis.* 1, *Southern limit of the wolves;* 2, *Northern limit of the jackals;* 3, *Distribution of the coyote.*

After J.P.Scott and J.L.Fuller

with local types of wolves, leading to the development of larger types of dogs, occur. The admixture in the dog of the hereditary material of many wolf types accounts for the great variability of the dog. Variability in type was further enhanced in domestication by the survival of types which in the wild state would have been eliminated.

Such a reconstruction is admittedly tenuous, but there are two additional considerations in its favor. One is the elimination of the jackal as ancestor of the wolf on cytological grounds. Thus all of Africa has been eliminated at one stroke as territory in which dogs might have been domesticated, for the wolf never reached there. While jackals cross readily with dogs in captivity, the dog widely found in Africa today must have been introduced as a domestic animal. The earliest evidence for dogs in Africa is a representation of leashed Saluki-like dogs. Later a variety of dogs were bred in Egypt, ranging from long-legged and short-haired greyhound or Saluki types with erect pointed ears to varieties resembling modern herding or hunting dogs. Even dachshund-like animals are shown on Egyptian frescoes. Whether the Saluki-like dogs were developed locally in Egypt or derived from imported stock cannot be determined on present evidence. Particularly strong trade relations between Mesopotamia and the Aegean [63] probably account for the appearance of Saluki-types in second millennium B.C. Cretan frescoes.

Similarly it appears that the dog was introduced as a domestic animal into North America, for if there had been independent domestication there, it is reasonable to suppose that the smaller canids of North America would have been domesticated, but none were. What probably did happen was that American wolves crossed with introduced Asian dogs.

GAZELLE, ANTELOPE, AND DEER. The effort to domesticate was not confined to the animals which have survived as domesticates to the present day. Gazelle, antelope, and deer were repeatedly domesticated, although in historical times such domestication efforts have been allowed to lapse. Remains of gazelle are abundant in Near Eastern archaeological deposits from the eighth to the fifth millennium B.C., and the animals were evidently much appreciated for meat. Presumably the successful domestication of other herd animals led to the domestication of gazelle and antelope. The difficulty of stalking the animals on the one hand, and the relative ease of taming them on the other may have been factors in their domestication.

Gazelle were eaten in Jericho beginning in the sixth millennium B.C., and it is possible that the animals were herded. While Mesopotamian seals from the Warka and Jemdet Nasr period (about 3200–2900 B.C.) show oryx antelopes, *Oryx dammah*, and gazelles in the context of temple herds, it is in Egypt that the most persistent effort at domestication occurred. It is certain that at least oryx, gazelle, and addax antelope were domes-

---

[63] Sasson, "North Syrian Economic Relations," pp. 177–79, and Canaanite Maritime Involvement in the Second Millennium B.C., *Journal of the American Oriental Society*, 86, No. 2 (April–June, 1966), 126–38. Zeuner, *Domesticated Animals*, p. 434.

ticated there by the middle of the third millennium B.C. Representations and tomb inscriptions show that they were as important in Egyptian animal husbandry as other herd animals, and that their use was as varied. Gazelle and antelope are included in the biblical list of animals permissible for food. Since the animals enumerated are overwhelmingly domestic, their inclusion supports the view that it was not before the end of the second millennium B.C. that gazelle and antelope began to disappear as important domesticates.

Fallow deer are shown confined in enclosures on Mesopotamian seal cylinders dating to ca. 2500 B.C. A hobbled deer serves as motif on a bridle ring found in a royal tomb at Kish in southern Mesopotamia which dates to ca. 2500 B.C. Since bridle rings generally depict domestic animals, it is possible that deer, too, were domesticated in the third millennium B.C. In Hittite areas the common fallow deer, *Dama dama,* was considered the sacral vehicle of a god and it may have been domesticated for religious purposes. There is some indication of its taming, if not of its actual domestication in a Hittite relief from Aliçar Hüyük where the animal is shown guided by rope and nose ring.[64]

Reindeer, *Rangifer tarandus*, is the only domesticated deer species of economic importance today. As Wolf Herre has shown, it is a relatively late domesticate, and clear proof of its domestication goes back only to the first millennium B.C. Reindeer were presumably a substitute domestication.[65]

STATUS DOMESTICATIONS. A number of animals were first domesticated to satisfy royal need for ostentation and ceremonial. The classification of such domesticates presents inevitable difficulties, for kingship was part of a larger sacral–social complex and it is impossible to identify ancient royal ceremonials, whether processions or hunts, which were not also sacral in character. This is not to say that animals thus domesticated were not used subsequently for purely secular and plebeian purposes. The importance of some animals, such as the cheetah and falcon, waned with the decline of sacral kingship, while other animals gained importance, as for example the elephant as a military juggernaut in India. The elephant's use in battle as well as in work and transport spread from India westward and also eastward to China, but where the first efforts at taming were made is uncertain. Elephants were common until at least the middle of the second millennium B.C. in the Palestinian–Syro–Mesopotamian area which remained, until the first millennium B.C., the center of ivory working.

Other status domesticates include lions, leopards, and tigers. None of these became truly domestic and supplies had to be replenished by repeatedly capturing and taming animals. These animals were trained to perform certain tasks, whether in war as the elephant, or in the hunt, like the cheetah and lion. Some animals were tamed because sanctity

---

[64] Brentjes, *Wildtier und Haustier,* p. 37.
[65] Herre, *Das Ren als Haustier.*

or high display value was attributed to them, as for example giraffes, crocodiles, apes, and monkeys.

BIRDS. Eduard Hahn regarded the goose as the oldest of all domesticated birds. Unfortunately, as Zeuner points out, the data do not allow one either to support or contradict this claim.[66] It is plausible that the domestication of the Nile goose, *Alopochen aegyptiacus*, dates back to the fourth millennium B.C., but firm evidence for the domestication of the gray lag, *Anser anser* and Chinese goose, *Cygnopsis cygnoides*, is available only from the end of the second millennium B.C. The domestication of the Egyptian Nile goose apparently did not lead to further domestications of geese and ducks for a considerable time. Indeed, Egyptian achievements in domestication had little impact on other African areas or on Asia. By the middle of the first millennium B.C. the Nile goose was displaced in the Egyptian food budget by introduced domestic fowl.

The migratory nature of geese and ducks probably retarded their domestication in the Near East, while the abundance of waterfowl in the southern Nile area made it possible to rely on hunting, which may have impeded the spread of bird domestication into the rest of Africa.

A great number of other birds were kept, tamed, and even domesticated at various times. Ostriches were kept in Egypt for their feathers and eggs, and were almost certainly bred in ancient Israel. Ducks may have been domesticated first in Mesopotamia, while pelicans, apparently highly valued for their eggs, may have been bred in Egypt. There is no evidence that they were ever used as fishing birds.[67] Whether cormorants were used for fishing in the eastern Mediterranean region where they were well known in ancient times, is not definitely known. The most widely accepted view is that they were first domesticated in Japan. Yet there are Mesopotamian representations of ca. 2500 B.C. which, according to Burchard Brentjes, point to the use of trained cormorant and osprey in fishing.[68] The widespread use of birds in the ancient Near East for purposes other than food makes it appear unlikely that the potential of fishing birds had been overlooked. Near Eastern societies that first utilized the homing instinct of pigeon and, as standard procedure, used ravens to assist navigators to find land [69] also might be expected to have exploited fishing birds. This seems all the more likely because it is clear that a variety of falcons, hawks, and eagles were trained for hunting in the Near East. These birds could not be bred in captivity and were therefore not domesticated in the strict sense of the word.

Falconry may have been developed as a royal sport and the Egyptian "Falcon of Horus" indicates an association of the bird with the sacral royal hunt. Royal falconry remained important in Mesopotamia at least until Asshur-ban-apal (668–626 B.C.).

[66] *Domesticated Animals*, p. 466.
[67] Zeuner, *Domesticated Animals*, p. 471.
[68] *Wildtier und Haustier*, p. 33.
[69] Feliks, *Animal World*, p. 88. The story of Noah in Genesis (5:7–12) reflects the ancient nautical use of both pigeon and raven.

AMERICAN ANIMAL DOMESTICATION. While plant domestications in the New World are both ancient and abundant, animal domesticates are comparatively few. Only the llama (*Lama glama*), alpaca (*L. pacos*), guinea pig (*Cavia porcellus*), and turkey (*Meleagris gallopavo*) were domesticated, in spite of the fact that a number of suitable animals were available. The best interpretation would seem to be that the Old World immigrants arrived without knowledge of animal domestication, although they probably did introduce the domestic dog from the Old World. Animal domestication in the New World, then, is either an independent achievement or the result of later contacts with the Old World.

The llama and alpaca were apparently first domesticated in Peru. Frederick E. Zeuner argues that the domestication of the guanaco, *L. guanicoë*, which gave rise to llama and alpaca was "certainly entirely independent of the domestication of Old World species." [70] It should, however, be kept in mind that evidence of Pacific connections for northwest South American cultures is particularly convincing. [71] Furthermore, it has now been shown that formerly llama were not confined to the cold steppes of the highlands of Equador, Peru, Bolivia, and northwestern Argentina, but were also found in the arid coastal areas of Peru. Art and osteological remains indicate that the animals were herded there and not, as was formerly believed, only imported from the highlands to supply meat for coastal populations.

Thus the possibility of an Old World stimulus for the domestication of llama and alpaca cannot be ignored. The llama, which is predominantly a beast of burden, and the alpaca, which is raised mainly for wool, have not spread beyond the range of wild ancestors. They are similar in this respect to a number of secondary Old World domesticates, especially south and southeast Asian bovines.

It is significant that the herding of llama has remained closely tied to sedentary farming and bears the stamp of recent domestication. In contrast to the Old World, no pastoral nomadism has developed, no ownership marks are in use, no leather or skin clothing is worn by owners or herders, nor are milk or blood of the animals used. There are no tents—in fact none of the equipment normally associated with pastoral peoples has been developed. On the contrary, pottery is widely used and, in a departure from Old World nomadic custom, meat is not fried or roasted. [72]

The extraordinary abundance of Old World animal domestications, including the many experimental, status etc. domestications, and the paucity of animal domesticates in the New World raise the question of why domestication of animals, or indeed of plants, should have been undertaken at all.

[70] *Domesticated Animals*, p. 436.
[71] See for example Emilio Estrada, Betty J. Meggers, and Clifford Evans, "Possible Transpacific Contact on the Coast of Equador," *Science* 135, No. 3501 (February 1962), 371–72.
[72] Horst Nachtigall, "Probleme des Indianischen Grossviehzüchtertums," *Anthropos*, 60 (1965), 177–97.

CHAPTER 6 *why were plants and animals domesticated?*

We have already touched upon the question of why plants and animals should have been domesticated in previous chapters. Here we shall focus more directly on this problem. In view of the tremendous importance that domestication has had for the development of civilization, the question may seem foolish. But the results of basic human inventions can rarely be foreseen: In some cases, those who first domesticated may not have considered the economic function that the domesticate came to fulfill.

Some animal domestication may have come about spontaneously. The ancestors of domestic dogs, pigs, and pigeons, for example, were probably "psychologically preadapted" for domestication. Wild pigs do not avoid human settlement, and it may well be that both the dog and pig initially sought out man, with man only gradually assuming the leadership in the relationship. Wild piglets are easily raised in captivity, and their behavior contrasts strongly with the ferocity of the wild boar, revealing "an amazing and unexpected plasticity in the personality of the pig." [1] This psychological plasticity is such that, conversely, domestic pigs lapse rapidly into wildness, becoming feral. On the other hand, there are species similarly predisposed to "self-domestication" that man has not seen fit to domesticate. These include birds like ravens and vultures, which are at least as obtrusive and forward as pigeons and yet were never domesticated. Bison, African buffalo, gazelle, and antelope were also domesticated in the past, while others as suited as these were apparently never considered for domestication.

Human psychology may of course be as important as animal predisposition to domestication. Domestication may have been furthered by instincts which make us cherish our own infants and are aroused

[1] Reed, "Archaeological Evidence," p. 139.

by young animals of somewhat similar bodily proportions. Piglets and dog pups are nursed by women in some primitive societies. Pigs raised from infancy form herds and remain in and near the settlement when grown, providing a ready expansion of meat supply. In Malaysia and New Guinea sows kept in such a semidomestic state cross freely with wild boars.

But psychological adaptability alone is not sufficient explanation. With sheep and goats the at best partial relevance of psychological explanations for domestication becomes evident. Animal breeders today find it difficult to breed enclosed wild sheep or goats, although captured wild goats become accustomed to man. Presumably, then, a considerable effort was involved in the original domestication of sheep and goats. Assuming that the exploitation of the animals involved first meat and hide, their relatively small size allowed a more economical use of the herd than would have been the case with larger animals such as cattle. The "meat and hide" argument, of course, does not address the question of why the animals should have been domesticated in the first place, when supplies of wild animals were ample, and hitherto fully used. Admitting that sheep and goats are earlier in the deposits than cattle, it remains uncertain whether sheep or goats were domesticated first. In any event, the questions that their domestication raise are similar, and indeed there seems to be a logical nexus in the domestication of the two animals.

Since the animals serve such similar purposes as domesticates, it might well be asked why, once either one was domesticated, men should have bothered to domesticate the other. Milk yield is low in the wild state in both animals, and thus, although goats have become important milk producers, this could not have been a consideration in their original domestication. In ancient art milking is shown in connection with cattle long before it is shown for goats. Wild sheep and goats have a similar coat, so that there was nothing to choose between the animals so far as their hides, hair, or wool were concerned. In goats real hairiness is a result of domestication: those with scimitar shaped horns, which are presumably closest to the wild form, have the least hair. Wild sheep have hairy kemp coats just like goats, and both have a wooly underfur, the degree of wooliness depending on the age, season, and subspecies. Wild goats would provide a hide just as warm as that of wild sheep, and if goats were domesticated first, the likelihood of the domestication of sheep would have been reduced. Wool coats in sheep are distinct features of domestication. The coat is the result of the retention of lamb coats by the adult animal,[2] as adult or adult-looking animals in the herds are killed, resulting in inadvertent selection for infantile characteristics in animals that were allowed to survive.

---

[2] J. E. Duerden, "Evolution in the Fleece of Sheep," *South African Journal of Science,* XXIV (December, 1927), 388–415, and J. E. Duerden and P. M. Seale, "The Ancestral Covering of Sheep," *South African Journal of Science,* XXIV (December 1927), 421–22. M. L. Ryder, "The Exploitation of Animals by Man," *Advancement of Science,* 23, No. 107 (May, 1966), esp. 11–13, 15. See further references on p. 18 of this article.

Should sheep be the older domesticate it might be supposed that the development of a wooly coat in the mature domestic sheep encouraged the attempt to domesticate goats in the hope that they too might develop into wool bearing animals. On the other hand, it is entirely possible that sheep and goats were first domesticated in one movement, that the fact that the animals are so similar and that their wild range largely coincides, meant that whatever motives led to the domestication of one operated concurrently in the domestication of the other. The animals may bave been treated as interchangeable in the earliest domesticating centers. Early Mesopotamian art makes no effort to distinguish between the two; by the time the animals had reached Egypt the distinction in art is much clearer. Moreover there seems to have been a conscious effort to equate the two and to breed animals that looked alike. This practice is still characteristic of India and East Africa where "the two species are often so much alike that they can be distinguished on close inspection only."[3] Finally, the widespread use of sheep and goats in religious ceremonial opens up the possibility that they were originally domesticated to ensure a permanent supply of animals for ritual, and economic uses were incidental.

The domestication of cattle raises many of the same questions that have been encountered in the discussion of sheep and goats, especially because this achievement cannot be explained as the result of an inadvertent process.[4] The size and fierceness of the wild ancestor required a strong motivation in the original domesticators for overcoming the difficulties of the task. Young animals might have been easier to capture than adults, but the problem of providing milk would have been insurmountable unless mother and calf were captured jointly or a gravid female was captured and successfully maintained in captivity. Thus the older nineteenth century theory that cattle were domesticated by being corralled for food, a theory still widely held today, raises many questions. It fails to explain the choice of certain animals and the rejection of others that were equally abundant, more easily captured, and more easily raised in captivity. Eduard Hahn argued, therefore, that the urus was domesticated for religious, not for economic reasons. He postulated that the motive for capturing and maintaining the urus in the captive state was to have available a supply, for sacrificial purposes, of the animal sacred to the lunar goddess worshipped over an immense area of the ancient world. Why the urus was selected as the animal sacred to the deity is uncertain, but probably because its gigantic curved horns resembled the lunar crescent. The bovine was early regarded as

[3] Zeuner, *Domesticated Animals*, p. 136.

[4] The following discussion rests mainly on Erich Isaac, "Myths, Cults and Livestock Breeding," *Diogenes*, 41 (Spring 1963), 70–93, and "Religious Factors in the Geography of Animal Husbandry," *Diogenes*, 44 (Winter 1963), 59–80. Diverging views based on a "human ecological" viewpoint can be found in *Man, Culture, and Animals: The Role of Animals in Human Ecological Adjustments*, Publication No. 78 of the American Association for the Advancement of Science, eds. Anthony Leeds and Andrew P. Vayda (Washington, D.C.: American Association for the Advancement of Science, 1965).

an epiphany of the goddess or her consort and was slain in the ritual reenactment of the myth of her/his death.

If cattle were domesticated because the horns of the urus resembled the moon's crescent, it is possible that other horned animals were also domesticated for their scimitar- or crescent-shaped horns, or for other reasons connected with ritual. Milking, if that is the case, may have begun as an attempt to relieve the mother animal after her lamb or kid was taken away for sacrifice. [It remains a historical problem why the goat alone became a milk animal on a large scale, the milking of sheep remaining a comparatively rare and localized practice (e.g., France). Presumably if selection had been made for milk yield, sheep could have become as high yielding milk animals as goats.] It also is possible that an unsuccessful attempt to domesticate crescent-horned gazelles, an attempt suggested by finds in the archaeological record, was made for ritual reasons. On the other hand, bison, which are domesticable but lack crescent-shaped horns, were never domesticated. A supporting argument for this approach is the extraordinary ceremonial importance of horns in ancient and medieval ceremonial and the preoccupation of many contemporary primitive herding people with horn configuration, a preoccupation which students of native animal husbandry in widely dispersed areas know is often detrimental to the economic performance of the breeds.

Hahn's followers have assumed that the process by which the urus was transformed into a domestic animal was as follows: the captured animals were kept in enclosed meadows for sacrificial use. Types different from the original strains of captured urus developed, for protected from predators and free to multiply, the sacrificial stock would have been either more or less inbred than under natural conditions. As every zoo keeper knows, this factor alone would produce deviations from the wild parent stock. Obviously, animals with more infantile characteristics, such as foreshortened heads, long legs, and relatively straight backs, as against the high withers and massive build of the wild cattle, could grow to maturity under the protective conditions of the sacred corral. Indeed, the selection of mature long-horned animals as epiphanies of the deity and thus the best animals for sacrificial purposes may have initially encouraged the survival of more infantile appearing individuals. Moreover, pied coats, which occur among many species as the result of domestication, developed in cattle as the result of breeding in confinement.

There is ample ethnological evidence that great ritual importance is attached to the marking and colors of animals, and the desire to produce more animals with a particular type of color pattern must have led to directional breeding. Thus, the argument runs, *Bos taurus longifrons*, an infantile appearing strain of sacred cattle more tractable than the parent stock, emerged. This made it possible to use cattle for ritual purposes other than sacrifice. Representations indicate that the first harnessing of cattle was to sleighs and wagons in religious processions. Ceremonial vehicles would subsequently have been modified for profane use. The plow, which from its earliest development appears to have been associated with cattle in ritual usage, would develop into the most characteristic tool of the "secular" agricultural system that originated in western Asia. Thus the argument for ritual domestication explains the origin of longifrons and at the same time the perpetuation of a urus-like animal, massive and long-horned, for ritual purposes. The sacred *Bos taurus primigenius* herds of the ancient Near East were the chief representatives of the massive animal.

Although primigenius cattle are the first domesticated type in the Near East and adjacent west Asia, in the earliest domestic cattle finds outside of these regions longifrons generally precede primigenius. The explanation is presumably that the tameness of longifrons as well as its early economic importance led to a rapid spread via trade and migration, while the difficulty of moving primigenius over long distances retarded its dispersal. The earlier presence of longifrons in North Africa, Crete; southeastern, Alpine, and central Europe, and southern Russia is strong evidence that neither was domestication of cattle achieved independently outside west Asia (although wild urus was common throughout these areas), nor was it simply the technique of domestication that was transmitted, for in that event, too, primigenius, the oldest domestic cattle, should be found first.

Since Eduard Hahn, a number of scholars have pointed to the close linkage between the domestication of cattle and a complex of religious ideas. The most convincing support for a religious motivation in the domestication of cattle is that the symbolic expression of the religious world picture with which this domestication was specifically associated has survived in areas far from the hearth of both the symbols and domestic cattle. In other words, migrations took place in which peoples accompanied by domestic cattle moved from the Near Eastern center of domestication to remote regions of Asia, Europe, and Africa. Other peoples in contact with the migrants learned from them and perhaps contributed in return to the migrating cultures. But the power of the "new" world picture was such that it became dominant; older conceptions were reinterpreted to harmonize with it; and the iconography of the intrusive cultivators and husbandrymen was accepted together with their technological achievements.

The religious revolution that I believe underlies agriculture supplanted the widespread Upper Paleolithic Eurasian religion, which centered as far as can be determined on mother goddesses with pronounced lunar and sometimes stellar attributes. The new world view rested on the belief that the world emerged from the death of a primeval being or god, whose slain and severed body came to constitute the existing world. In its most ancient form the myth involved the death and resurrection of a mother goddess. In the Mesopotamian Gilgamesh epos, for instance, the hero, with his friend Enkidu, kills a monster Humbaba, an equivalent of the Hurrian mother goddess. This is one version of a world creation myth; there are many others, as for example the myth of Tiamat, a primeval serpent from whose cloven body heaven and earth are made. In the new world picture the goddess slowly becomes part of a triad or a divine family in which often the consort or son usurps the role of the slain deity.

In the transitional period of the early mythological cycles of the Near East the sex of the sacrificed deity is blurred. Thus the goddess Astarte has a more ancient forebear, Athtar, who appears as a male god and, to judge from Acadian evidence, was an androgynous deity. In Sumer, too, a god might be addressed in successive lines of a very early hymn as "mighty young bull . . . fruit which begets itself . . . womb which bears everything." The recognition of the fundamental importance of sex distinctions, essential to the development of agriculture and animal husbandry, is implicit in the new divine families and in the basic myths of Near Eastern religion—the myths of the slain and resurrected fertility gods. At the same time the myths refer predominantly not to the creation of the world, but to the emergence of plant life from the slain body. The myth has become the distinctive myth of agriculture. The earliest formulation is in the Ishtar (Inanna)-Tammuz cycle, but the cycle is omnipresent. In the Canaanite Baal cycle, the goddess is the slayer:

> She seizes the Godly Mot—
>   With the sword she doth cleave him
> With the fan she doth winnow him—
>   With fire she doth burn him
> With hand-mill she grinds him—
>   In the field she doth sow him [4a]

All the myths of the dying fertility gods are ritual myths, which means that they express what is conceived as the key to a culture's origin, meaning, and survival. Hence they must be reenacted at critical times, such as a change of season, crisis periods of human life like death and birth, and in national crises. It is the ritual which shores up and defends, in fact reestablishes the order of existence as it was originally constituted before the threat to that order. It is the fact of ritual reenactment which indicates that a particular narrative is a "true" myth and not merely a fanciful tale or legend borrowed from somewhere else.

Not only was the myth of the slain and reborn god a ritually reenacted myth, but the ritual itself shows so many detailed correspondences in far flung areas that a multiple independent development of either ritual or myth is most unlikely. Thus an original connection between ancient agricultural systems would be supported, even if there were no other evidence such as that concerning the origin and heredity of domestic cattle.

The domestication, certainly of cattle, and perhaps of the other horned animals, sheep and goats, can be understood in the context of the myths. Crescent-horned animals, an epiphany for both the lunar

[4a] H. L. Ginsberg, trans., in James Pritchard, ed., *The Ancient Near East: An Anthology of Texts and Pictures* (Princeton, New Jersey: Princeton University Press, 1969), pp. 112–13.

goddess and her consort, became a favorite substitute for the god in the reenactment of the myths, and the necessity for a permanent supply of sacrificial stock led to cattle domestication. It must be understood that the sacrifice was not primarily the offering which we now interpret sacrifice to be, but an actual reenactment of the original sacred happening—in this case, the death of the god.

The lunar association of the older mother goddesses and then of the slain gods in the fertility myths reflects what must have been the great discovery that the cycle of the moon coincided with the human female fertility cycle. In any event, wherever the ritual cycle of the dying god is found, it is accompanied by a well-known strong association with cattle as one of the deity's main epiphanies. Moreover, the castration of cattle, which is surely one of the great agricultural innovations, resulting in the ox, can only be understood as a by-product of the role of cattle in the personification of the divine fate. Neither the docility of oxen nor improved meat texture of castrates could have been foreseen so that the results of castration can not be cited as the reason for the practice. Human ritual castration probably served as model for the castration of animals. In many of the rites of the suffering gods, human castration enacted the mythical self-mutilation of the gods described in the Agdistis-Attis, El and Kronos-Uranos myths. The castrated worshipper of Attis became one with Attis and was, in fact, called Attis. Eunuchs played a leading role in the cults of Bronze Age Syria and Anatolia, and in certain periods in Mesopotamia itself. In Cappadocia, Assyrian tablets of the nineteenth century B.C. use the name "Kumrum" (eunuch) as the conventional title for priests. Animal castration followed the human model and was as much a sacral performance in *imitatio dei* as the former.

Hahn's thesis of a religious motive in the domestication of cattle raises a number of far reaching questions. First, how did the world picture that led to the domestication of cattle arise; and second, does this world picture not reflect an intellectual or religious revolution which underlies the entire movement toward plant and animal domestication? Concerning the first question, there is no way of knowing what the observational, logical, or traditional foundations of the myths of the slain and resurrected deities were: perhaps simply the observation of seasonal vegetation cycles was translated to mean that death is a condition of birth. This conclusion may have been reached through the additional experience that from decaying parts of an organism new life forms emerge—often apparently very different from the life that perished. Thus since dead plants, animals, and men seemingly give rise to other forms of life (a belief, incidentally, that was widely held and not abandoned until long after the beginnings of modern biology) the conclusion that death was at the beginning of all life appears a comprehensible one. But if the first death produced the world order and its living things, it followed that life and the world's continuation could only be safeguarded by the ritual repetition of the first death.

To answer our second question, it is possible that cultivation and domestication as a whole were the outcome of this world picture. For if death is at the root of existence, ritual repetition of the first death may not merely safeguard life but actually increase it. To put it crudely, the "insight" that "death preceded life or rebirth" could have led to the conclusion that killing will multiply life.

Assuming that the mythical perspective of the domesticating cultures is correctly interpreted, then the imperatives of ritual reenactment could have proven a tremendous innovating force, leading to a kind of ritual experimentation in which plants were "slain," i.e., cut up and buried. It would seem that particular plants may have been considered most suitable, and were closely tied into the ritual. Similarly, appropriate animal epiphanies were attached to the mythological-ritual performances by a process of accretion and reinforcement, whose socio-psychological foundations are only partially understood. In any event cultivation and the domestication of plants and animals were the byproduct or outcome of a religious world picture. The trend of this argument is to reverse the popular Marxist axiom that religion and science are superstructures. For in this case technology (domestication) was a superstructure on or at least an integral part of religious knowledge.

The thesis of the religious root of domestication has been exploited further to support a broad geographic regionalization of the primary hearth of domestication. The ritual myth is found both in the ancient Near East and in areas of tropical agriculture. An independent origin of the myth can be dismissed. Presumably lunar fertility myths could arise in different areas independently, for the moon's behavior is everywhere observably regular and can be connected with human fertility. But it seems much more unlikely that the same observation should independently give rise to very similar rites. If the ritual myths of the slain god are linked, did they originate in the tropics or outside them? If the area of origin could be found, that discovery would provide an important reason for supposing that the origin of agriculture was in the same place.

The ethnologist A. Jensen shows that primitive tropical cultivators dramatize in their various rites a central myth which describes the origin of food crops as stemming from the murder of a lunar maid or goddess.[5] Jensen's theory is based largely on southeast Asian data and he is inclined to accept therefore a tropical origin of agriculture. Of course anthropologists in general have tended to identify tropical cultivation as the oldest form of agriculture. The absence of archaeological support for this view is not considered pertinent, and tropical root and tuber cultivation is either considered to be ancestral to grain agriculture or to have developed independently of it. In a recent study Jensen noted that in tropical areas, where native cultivators plant tubers and root crops *and* cultivate grains, the myth of the dying god is associated with the former. A different myth of origin is generally attached to grains, one that describes them as brought from heaven, usually against the will of the deity. Jensen distinguishes between the two mythical motifs, calling

[5] A. E. Jensen, *Das religiöse Weltbild einer früher Kultur,* Studien zur Kulturkunde bd. 9 (Stuttgart: August Schröder Verlag, 1949).

one the "Hainuwele motif" after the Indonesian version, and the other the "Prometheus motif." [6] The Prometheus myth refers to the theft of grain and the Hainuwele motif to the myth of the slain being whose body gave rise to tropical root crops.

Jensen finds that a sharp division is made between the two crops in practice as well as in myth. The Darassa in Ethiopia store grain flour under the aegis of the men and flour from root crops under the protection of women. Where tuberous plant crops accidentally produce seed, the resulting crop may not be used but must be destroyed. Jensen points out that in Ethiopia no tribe that had a myth of agricultural origins failed to have a separate myth for the two types of plants. Jensen notes that outside of the large African area where the dual myths are found, they are sometimes reversed. In some areas, for example, the Prometheus myth may be attached to tubers and the Hainuwele myth to rice. But on the whole even Indonesia fits the pattern. In the western areas, where rice and millet predominate, the Prometheus motif is common; in the central islands where root and tree crops are most important, the "Hainuwele motif" is the myth of agricultural origin. In still other areas of Indonesia both kinds of crops are held to have sprung from a primeval being, or both are held to have been stolen from heaven. These are regarded by Jensen as reversals that may be expected in the process of diffusion of the myths. In any event Jensen believes that the two categories of myths originally applied by tropical cultivators to grain and tuberous domesticates makes it unlikely that tropical agriculture could have derived from a primary west Asian hearth of agriculture. He agrees with the common ethnological view, considering it much more likely that tropical agriculture is older or at least that the two agricultural complexes originated independently.[7]

Nonetheless it can be argued that the two types of myths actually point to a Near Eastern hearth which preceded and influenced the development of tropical cultivation, for the tropics have only the modified and secondary form of the myth, that which is applied only to the origin of domestic plants. The Near Eastern myth started as a world creation myth and was only subsequently specialized to become the myth of agriculture. It would seem that the area which has both the most comprehensive form of the myth and the specialized version stemming from it is the hearth of the myth, rather than an area which has only the specialized secondary version. Indeed the Prometheus myth itself suggests a foreign origin for the crops whose original cultivation it explains, for in the myth grain is taken or stolen from its original owners. As a ritual myth, Jensen's Hainuwele myth lends itself both to grain and root crop cultivation. It is adapted to root crops because the slaying of the body (cutting of plants to set out cuttings) leads to new and multiplied life. It is also appropriate for cereals, for the operations of harvesting "slay" the plant. Moreover, as the preceding chapters indicate,

[6] *Ibid.*, 34–40, and A. E. Jensen, "Der Ursprung des Bodenbaus in mythologischer Sicht," *Paideuma*, VI, No. 3 (April 1956), 169–80.
[7] Jensen, "Ursprung des Bodenbaus," p. 176.

there is no independent evidence supporting the derivation of the Near Eastern hearth from a prior tropical center or centers. That the flow of the myth, therefore, was from the tropics to the Near East, would also seem highly unlikely.

While it can never be more than an hypothesis that the origin of domestication was religious, there are cases of domestication of specific animals and plants where the religious motivation seems quite clear. The pigeon, for example, would seem to be a classical case of the exploitation of a symbiotic tendency, for it is essentially a self-domesticating bird which seeks out human fields and settlements. But the self-domestication was probably facilitated by the extraordinary role the pigeon played in the religious conceptions of early Near Eastern farming cultures. Early in the fourth millennium B.C. pigeons were closely associated with ritual. They became symbols of the ancient mother-goddess and the symbolism spread with the migration of farming cultures. In Christian inconography the mother goddess and her symbols are still present and the pigeon is a symbol of the Holy Ghost. The pigeon, or dove, continues to be the symbol of peace in secular ideological movements. The economic exploitation of pigeons, too, is still hedged in many areas by the old awe and love shown to the bird. It is not eaten by Ethiopian Christians and eaten very rarely in most parts of the Islamic world. This attitude parallels and has the same ancient sacral roots as the food avoidance of the dove by Orthodox Christians in Tsarist Russia.

Domestic fowl, today the most economically important domestic bird, was probably first domesticated for religious reasons. Carl Sauer believes that no economic motive entered into their domestication,[8] for it is clear that egg laying and meat producing qualities are relatively late consequences of their domestication. The religious intent in fowl keeping may be connected with cock-fighting, a ritual reenactment of a mythological divine combat. Fighting cocks were also used for oracular purposes. Similar attempts at ritual domestication were undertaken with quail (*Coturnix coturnix*) in ancient Greece. Quail cocks, like roosters, were matched in combat, but the effort at ritual domestication of quail failed.[9]

Fowl appear in a number of varieties in Indus valley art of the third millennium B.C., in Persia in the third millennium B.C., and elsewhere in the Near East in the second millennium B.C.[10] The waking crow of the cock is closely associated with solar rituals of western Asia and fowl became an integral part of the Near Eastern solar iconography of the second millennium B.C.[11] Indeed the mythology of the "fire cock" has spread

---

[8] *Agricultural Origins*, p. 32.
[9] Zeuner, *Animal Domestication*, p. 458.
[10] *Ibid.*, pp. 443–55.
[11] The Talmud with anti-paganizing intent stipulates: "When a man hears the crow of the cock he should say: 'Blessed be the One who has given the cock understanding to distinguish between the day and the night.'" This blessing was included in the Jewish morning prayer.

widely and survived in the folklore of western Eurasia. The poultry egg was recognized as symbol of death (and rebirth) at least as early as the beginning of the first millennium B.C.[12]

But the clearest case of religious motivation in the domestication of an animal is the cat. Near Zagazig a complete record of mummified cats exists in Egypt (the cat was the epiphany of the goddess Bast), showing a sequence of development from the ancestral Libyan wild cat *Felis catus Libya*, to the domestic cat.[13] Domestic forms appear first in the course of the twelfth dynasty, ca. 2000 B.C. Nowhere outside of Egypt were wild cats domesticated. The famed ethologist Konrad Z. Lorenz has explained the domestication of the cat by suggesting that the Libyan wild cat entered into some kind of symbiotic relationship with ancient grain-storing Egyptians.[14] But this suggestion is not convincing, for even today there is a symbiotic relationship between the wild cat and man in southern Nubia and, despite the long history of this relationship, no development toward domesticity has taken place.

Other rodent hunters were used effectively in the ancient period, including the ferret and mongoose. In Egypt, the Mediterranean area, and most of Europe widespread use was also made of the "house snake," and cats did not displace them in many areas until the Christian era. The cat has not been identified in the Bible, but it is interesting that the postbiblical Hebrew term for cat is *hatul*, "the swaddled," undoubtedly referring to the tradition of cat mummification in Egypt. Moreover, it is likely that our word "cat" ultimately derives from the old Semitic word for flax and linen, from which our "cotton" also derives. In other words, the term "cat" refers to the material in which the cat mummy was wrapped. The cat is found in the Germanic culture area as sacred to Freya and has of course survived in folklore as the familiar of the witch.

Religious motives also seem clear in the domestication of some plants. The citron, for example, may have been the first of the citrus trees domesticated in Asia, and was taken over by ancient Israel, propagated, and dispersed because of its importance in Jewish ritual.[15] Many of the dye plants were brought into domestication for their ceremonial value. It may be that the origins of clothing are not in any need for protection against sun and weather, but were for social classification, religio-magical purposes, or a combination of these.

Tattooing and scarification as well as painting of the body of the living and the dead with vegetable and mineral body paints would seem explicable only in terms of psychological needs. Thus the domesti-

[12] Brentjes, *Wildtier und Haustier*, p. 33.
[13] Bodenheimer, *Animal Life*, II, 372–75.
[14] *Man Meets Dog* (London: Pan Books Ltd., 1959), pp. 22–24.
[15] Erich Isaac, "Influence of Religion on the Spread of Citrus," *Science*, 129, No. 3343 (January 23, 1959), 179–86, and "The Citron in the Mediterranean: A Study in Religious Influences," *Economic Geography*, 35, No. 1 (January 1959), 71–78.

cation of turmeric may well have been a religious domestication.[16] Of course this plant is important in various folk-pharmacopeias and it might be argued that its medicinal value encouraged its domestication. A number of other plants that may belong to the oldest cultigens of southern Asia were perhaps also domesticated ritually. These include cordyline, ginger, gandarusa, and holy basil.[17]

It is interesting to note, however, that many plants with distinct physiological and psychological effects on man have not been domesticated.[18] In many instances systematic cultivation is only the result of contemporary search for new drugs. The reason is probably that obtaining the plant was part of the magical curative process and a plant that was cultivated would not be endowed with the same potency as the rare plant growing in a fittingly wild setting. Magic and medicine have been inextricable until quite modern times.

The role of religion in the dispersal of plants is better known than its role in their domestication, for necessarily there is more evidence concerning more recent geographic dispersals than concerning ancient origins. The case of the citron was just mentioned. The sacral value of wine led to the introduction of grapes by Jews and Christians into Europe north of the Mediterranean littoral. The religious factor has also been important in the dispersal of plants without economic significance. Thus the Indian pipal tree has been introduced by Buddhists into Ceylon and Japan.[19] It is possible that Islam accounts for the dispersal of the white iris of Yemen. The Yemen white iris, *Iris allicans,* was probably introduced into Spain, the Canary Islands, France, Corsica, Cyprus, and Turkey because of the high regard shown to the flower as a Muslim cemetery plant. The blue iris of Yemen, not used in this way, did not spread.[20]

An intriguing question is whether the affinities between the world pictures of pre-Columbian America and Old World domesticators can be explained by actual migration of Old World peoples to the Americas —perhaps before they had domesticated plants and animals, but after

---

[16] David E. Sopher, "Indigenous Uses of Turmeric (*Curcuma domestica*) in Asia and Oceania," *Anthropos,* 49 (1964), 93–127, and Joachim Sterly, "Gelbwurz (*Curcuma* spp.) als Ritual und Heilmittel in Melanesien," *Anthropos,* 62 (1967), 239–40.

[17] J. E. Spencer, *Shifting Cultivation in Southeastern Asia,* University of California Publications in Geography, No. 19 (Berkeley and Los Angeles: University of California Press, 1966), pp. 114, 154.

[18] Cf. Margaret B. Kreig, *Green Medicine: The Search for Plants that Heal* (Chicago: Rand McNally, 1965), and Clyde M. Christensen, *The Molds and Man,* third revised edition (Minneapolis: University of Minnesota Press, 1965), especially pp. 183–225. See also Gertraude Freudenberg and R. Caesar, *Arzneipflanzen* (Berlin: Verlag Paul Parey, 1954).

[19] David E. Sopher, *Geography of Religions,* Foundation of Cultural Geography Series (Englewood Cliffs, N.J.: Prentice-Hall, Inc., 1967), p. 33.

[20] F. C. Stern, "Plant Distribution in the Northern Hemisphere," *Geographical Journal,* 108, No. 1–3 (July-September, 1946), 24–40. See especially p. 26. It should be noted however that already the *Mishna* (second century A.D.) mentions this flower as commonly found in cemeteries (*Tohoroth* III:7), and the Midrash (*Vayikra Rabba* 23:5) describes it as a flower "whose heart is directed to heaven." It is possible that Islam accepted this symbolism from the Jews. In Christian Europe the main cemetery flower became the lily. Cf. the German *Leichenlilie.*

the new religious conceptions had developed. There are some striking resemblances between Old and New World rituals of agriculture. Among the Aztecs there was a festival of "the killing of the god Uitzilopochtli" in which the image of the god, fashioned in dough and fortified with human blood, was "slain" by a priest in the temple. The effigy was then cut into pieces and consumed by all the men of the community. There were a number of agricultural festivals in which similar rites were performed. All the elements of the myths of the slain gods in the Old World agricultural societies are illustrated here: the death of the god, the sanctioning of food through that god's death, and the establishment of a world order as a result of the god's death in which man and plant share one fate.[21]

The stress on the irrational springs of domestication is not meant to imply any general view that all technological developments are superstructures on religious world views. Even in the case of domestication, the secondary and substitute domestications were generally undertaken for economic reasons; and of course the primary domesticates as well soon served economic functions primarily and their religious meaning was in some cases forgotten or became secondary. What is argued here is that the initial domestication of plants and animals constituted such an enormous break with the past that an intellectual revolution must have preceded the economic development, and that there is evidence for such an intellectual and inevitably religious revolution.

The categories which we apply, including "religious" versus "economic" or other motivations, are modern categories of thought and cannot be applied in any simple way to the men who first domesticated. Even contemporary man does not differentiate completely between technical, utilitarian, aesthetic, and religious aspects of the objects in his environment. Durkheim has noted that religion is

like the womb from which come all the leading germs of human civilization. Since it has been made to embrace all of reality, the physical world as well as the moral one, the forces that move bodies as well as those that move minds have been conceived in a religious form. That is how the most diverse methods and practices, both those that make possible the continuation of the moral life (law, morals, beaux-arts) and those serving material life (the natural, technical and practical sciences), are either directly or indirectly derived from religion.[22]

[21] E. O. James, *The Beginnings of Religion,* Arrow Books (London, Hutchinson & Co. Publishers, Limited, 1958), 92–93. The rites are described by James who uses J. de Acosta's *Natural and Moral History of the Indies.* Striking parallels are known from the Old World, cf. A. M. Blackman, "Osiris as the Maker of Corn in a text of the Ptolemaic Period," *Studia Aegyptiaca* I–II (Rome: Pontificium Institutum Biblicum, 1938), and Henri Frankfort, *Kingship and the Gods* (Chicago: University of Chicago Press, 1948), 86–127 and *passim.* The parallels in the ritual reenactment of the Old and New World mythologies argue against a fortuitous coincidence of "archetypal forms." The archetypal and visionary origins of Mexican religions has been recently suggested again by C. A. Burland, *The Gods of Mexico* (New York: G. P. Putnam's Sons, 1967), 127–30.
[22] Emile Durkheim, *The Elementary Forms of the Religious Life,* trans. Joseph Ward Swain (New York: Collier Books, 1961), p. 255.

postscript

The state of knowledge concerning domestication, as this volume has shown, is incomplete, but knowledge of plant and animal domestication, and of the dispersal of domesticates, is rapidly being enlarged by advances in many fields. Biologists in particular have concerned themselves increasingly with the study of the heredity of domesticated plants and animals. While ostensibly the main purpose of this research is to increase knowledge of germ plasm resources that may be used in the development of new varieties and breeds, this work also yields information about the relatedness of different domesticated species.[1] In many instances agricultural genetics has also succeeded in determining the wild ancestor, or at least in narrowing the range of wild animals or plants which may be considered ancestral to a domesticated variety. At

[1] In addition to works previously cited on the genetics of cultivated plants, useful references are C. D. Darlington, *Chromosome Botany and the Origins of Cultivated Plants,* 2nd. ed. (New York: Hafner Publishing Co., 1963); C. D. Darlington and A. P. Wylie, *Chromosome Atlas of Flowering Plants,* 2nd. ed. (London: George Allen and Unwin Ltd., 1955). See also J. R. Harlan, "Geographic Origin of Plants Useful to Agriculture," in *Germ Plasm Resources,* ed. Ralph E. Hodgson, Publication No. 66 of the American Association for the Advancement of Science (Washington, D.C.: 1961), 3–19. A concise introduction to agricultural genetics is James L. Brewbaker's *Agricultural Genetics,* Foundations of Modern Genetics Series (Englewood Cliffs, N.J.: Prentice-Hall, Inc., 1964). Studies in serology and genetics may illuminate current taxonomic problems. On bovines, for example, see Clyde Storment and Yoshiko Suzuki, "The Distribution of Forssman Blood Factors in Individuals of Various Artiodactyl Species," *Journal of Immunology,* 81, No. 4 (October 1958), 276–84; Clyde Storment, "Current Status of Blood Groups in Cattle," *Annals of the New York Academy of Sciences,* 97, Article 1 (May 3, 1962), 251–68. P. W. Gregory and F. D. Carroll, "Evidence for the Same Dwarf Gene in Hereford, Aberdeen-Angus, and Certain Other Breeds of Cattle," *Journal of Heredity,* XLVII, No. 3 (May-June 1956), 107–11; P. W. Gregory, W. S. Tyler, and L. M. Julian, "Bovine Achondroplasia: The Reconstitution of the Dexter Components from Non-Dexter Stocks," *Growth,* 30 (1966), 393–418.

the same time advances in the study of the comparative anatomy of domestic and wild varieties have helped to establish the domestic or nondomestic status of plant or animal remains in archaeological deposits.

In addition, advances in archaeology and an increasingly dense net of prehistoric archaeological coverage of the world will help to clarify relationships between prehistoric domesticators and food gathering peoples. Finally, increasingly sophisticated approaches to the study of past climates and faunal and floral distribution pertinent to the ecology of prehistoric society may further our knowledge of the history of both plant and animal domestication.[2]

But what of future domestication? It is possible that modern man has entered into a new phase of domestication. The search for drugs with the most varied properties has called attention to wild plants used for curative or poisonous results by many aboriginal peoples. In some cases such plants are cultivated experimentally, but it is doubtful whether many domestic strains will be developed, for cultivation will presumably be abandoned once the active ingredients are synthesized. New strains of molds and fungi, for example, have been produced for medical and other beneficial research purposes, and it is to be expected that "domesticated" strains have been developed for purposes of war as well. The awful potentialities of this kind of domestication for the geography of man and his traditional domesticates are apparent.

Less forbidding is the development of domestic varieties of formerly gathered plants, e.g. various berries. The dangers of environmental deterioration and disease in cultivated plants has stimulated the interest of agricultural geneticists in their wild relatives. These constitute an as yet barely exploited resource of germ plasm which may be used to develop domestic varieties that can cope satisfactorily with various environmental contingencies. The effort to combat "Underdevelopment" has also encouraged a search for new crop plants in underdeveloped countries, but this effort is as yet in its infancy.

The interest in domesticating new animals today cannot compare with the interest shown in new plant domestications. No significant gains are expected from the domestication of new species. Occasional experimental domestications have been undertaken, for example, the Arctic musk oxen, African eland, antelope, zebras, and the European elk and bison. In no instance, however, have such domestications become economically successful. On the contrary, the animals are maintained only by private funds or subventions from the public purse. The value of such domestications consists in the demonstration that they are technically feasible and in the new knowledge of animal behavior. Economically such domestications have generally failed because of the high costs involved in inserting new animals into the established marketing systems of animal products and because of cultural resistance on the part of potential consumers. Nonetheless new animal domestications should be tried in suitable areas of the underdeveloped world, particularly in regions where environ-

[2] Cf. H. E. Wright, Jr., "Natural Environment of Early Food Production North of Mesopotamia," *Science*, 161, No. 3839 (26 July 1968), 334–38.

mental stresses make the introduction of modern animal husbandry resting on traditional domestic animals both technically difficult and expensive. It seems at this time to be feasible to develop in Africa, for example, an animal husbandry based on the herding of wild giraffe, elephant, hippopotamus and numerous varieties of antelope.[3] Whether such wild animal management will be successful and whether it will lead to domestication will depend ultimately on changes in the cultural perception of these animals by African societies.

Of the insects, only the bee, *Apis mellifera syn. A. mellifica,* has been subject to continuing efforts at domestication.[4] The transition from gathering honey to the possession of swarms is usually explained by referring to common practices among a variety of contemporary primitives. Food gathering peoples often provide hives to attract flying swarms. With the development of methods of harvesting the bee's products (honey, wax and propolis) without killing the swarm, the transition from mere destructive exploitation to domestication was accomplished. The insect had religious significance in the countries around the Red Sea and the eastern Mediterranean and it is possible that the domestication of bees may have been based on ritual needs.[5] It also is possible that bee keeping arose from the needs of warfare where the hives of bees were hurled against the enemy. This use of hives was known as late as the Thirty Years' War.[6] It is interesting in this connection that the effort of bee domestication was directed primarily to stinging bees, ignoring the species of tropical stingless bees (*Melipona* and *Trigona*). Modern apiculture is still engaged in an effort to diminish the danger while retaining the yield, for unfortunately the more productive bees are also among the most dangerous.[7]

[3] F. Fraser Darling, "Wildlife Husbandry in Africa," *Scientific American,* 203, No. 5 (November 1960), 123–34.

[4] A number of formerly domesticated insects have been given up. Sericulture has been overwhelmed by synthetic fibres and the Chinese silk moth, *Bombyx mori,* is going the way of the Near Eastern silk moth, *Pachypasa otus,* an abandoned domesticate. Synthetic dyes have obviated attempts to domesticate scale insects formerly harvested for dyes, e.g., the Mexican cochineal scale, *Coccus cacti,* or crimson, *Kermes ilicis* or *K. biblicus.* Insects are widely used for food, but none have been domesticated for that purpose. See F. S. Bodenheimer, *Insects as Human Food: A Chapter in the Ecology of Man* (The Hague: W. Junk, 1951).

[5] Zeuner, *History of Domesticated Animals,* 501–4. On the mythological connection of bees and cattle see Erich Isaac, "Relations between the Hebrew Bible and Africa," *Jewish Social Studies,* XXVI, No. 2 (April 1964), 87–98, especially pp. 95–98.

[6] Zeuner, *History of Domesticated Animals,* 504. Cf. Exodus 23:28: "And I will send the hornet before thee, which shall drive out the Hivite, the Canaanite, and the Hittite, from before thee." Wasps have apparently maintained a status formerly shared with bees as is indicated by a North Vietnamese Press Agency report that a South Vietnamese peasant "is readying hornets to attack U.S. and Saigon forces." William Beecher, "Way-Out Weapons," *The New York Times Magazine* (March 24, 1968), p. 54.

[7] The difficulties are illustrated by the abortive Brazilian effort to acclimatize African bees. "Though it is known to be ferocious, the African bee produces 30% more honey than either the Italian or German bee . . ." The results of the introduction were devastating: "Quick to anger, even quicker to swarm, the new Africans have turned on Italian and German bees . . . killing off hive after hive . . ." The only solution found by apiculturists is the destruction of the imported bees. "Destroy them all," says Father Nedel. "If they are not controlled . . . they will take over Brazil." *Time* (September 24, 1965), p. 75.

Strictly speaking the honey bee is a self-domesticating animal. Like the ant it evolved complex social structures including functional castes, sedentary settlements with controlled internal environments, hostile attitudes to other communities, the limitation of individual procreative activity, and total self-sacrifice for the sake of the community. Self-domestication in bees was accomplished about thirty million years ago, and since that time morphological and societal changes in the honey bee have been negligible. Zeuner believes that examples from the animal world such as the honey bee show that "in successful species, stabilization is invariably achieved." [8]

The most formidable future domesticate is perhaps man himself. A self-domesticating animal even now, shaped by his own cultures and social systems, man may create entirely new human organisms. Ectogenesis, propagation by test tube and artificial womb, is now in a state of advanced experimentation. Selection for sex followed by genetic surgery, the insertion of wanted genes and the destruction of undesirable ones, are reasonable forecasts for the next decades. The more distant prospect is for "cloning," the production of men genetically identical to the original donor. New life forms might be created by the combination of material from human and animal nuclei. Such forms might be created for the colonization of new areas—whether terrestrial, planetary, or in outer space—for which men are not biologically suitable.

We have lived through a period in which science fiction has become less and less fiction. It is in the area of sociological science fiction that reality has not caught up with fantasy. But the potential of genetic manipulation paves the way for changes in the structure of human societies as radical as those that technology has produced in material culture. Class systems might be based on biologically manipulated differences. There is the potential for a society as stable as that of the bees, with those who perform menial tasks biologically capable only of such tasks, and those who perform other tasks similarly genetically constructed to fulfil their roles. Huxley's *Brave New World* would seem more likely than Orwell's *1984*, at least for 2084.

The force of conservatism is greater where the manipulation of human biology is concerned, precisely because of the awesome implications. Yet the pressure of man's scientific achievements are such that it may prove impossible to resist such manipulation. Perhaps human societies as presently known cannot cope with their own destructive technological capacities. Biological engineering may be seized upon as a way, perhaps in the long run a deceptive one, to avert social catastrophe.

[8] Zeuner, *History of Domesticated Animals*, p. 508.

index